T0136415

'C' Programming in Open Source Paradigm: A Hands on Approach

RIVER PUBLISHERS SERIES IN INFORMATION SCIENCE AND TECHNOLOGY

Volume 20

Series Editors

K. C. CHEN
National Taiwan University
Taipei, Taiwan

SANDEEP SHUKLA
Virginia Tech
USA

CHRISTOPHE BOBDA
University of Arkansas
USA

The "River Publishers Series in Information Science and Technology" covers research which ushers the 21st Century into an Internet and multimedia era. Multimedia means the theory and application of filtering, coding, estimating, analyzing, detecting and recognizing, synthesizing, classifying, recording, and reproducing signals by digital and/or analog devices or techniques, while the scope of "signal" includes audio, video, speech, image, musical, multimedia, data/content, geophysical, sonar/radar, bio/medical, sensation, etc. Networking suggests transportation of such multimedia contents among nodes in communication and/or computer networks, to facilitate the ultimate Internet.

Theory, technologies, protocols and standards, applications/services, practice and implementation of wired/wireless networking are all within the scope of this series. Based on network and communication science, we further extend the scope for 21st Century life through the knowledge in robotics, machine learning, embedded systems, cognitive science, pattern recognition, quantum/biological/molecular computation and information processing, biology, ecology, social science and economics, user behaviors and interface, and applications to health and society advance.

Books published in the series include research monographs, edited volumes, handbooks and textbooks. The books provide professionals, researchers, educators, and advanced students in the field with an invaluable insight into the latest research and developments.

Topics covered in the series include, but are by no means restricted to the following:

- Communication/Computer Networking Technologies and Applications
- Queuing Theory
- Optimization
- Operation Research
- Stochastic Processes
- Information Theory
- Multimedia/Speech/Video Processing
- Computation and Information Processing
- Machine Intelligence
- Cognitive Science and Brian Science
- Embedded Systems
- Computer Architectures
- Reconfigurable Computing
- Cyber Security

For a list of other books in this series, visit www.riverpublishers.com

http://riverpublishers.com/series.php?msg=Information_Science_and_Technology

'C' Programming in Open Source Paradigm: A Hands on Approach

K. S. Oza
S. R. Patil
R. K. Kamat

Department of Computer Science,
Shivaji University,
Kolhapur,
India

Published, sold and distributed by:
River Publishers
Niels Jernes Vej 10
9220 Aalborg Ø
Denmark

River Publishers
Lange Geer 44
2611 PW Delft
The Netherlands

Tel.: +45369953197
www.riverpublishers.com

ISBN: 978-87-93237-67-4 (Hardback)
    978-87-93237-68-1 (Ebook)

©2015 River Publishers

Contents

Foreword

'C' language over the period of last few decades has become the icon of computer programmers. The field of computer science has undergone tremendous changes, and the rate of obsolescence of concepts, programming platforms, tools, and utilities is extremely high. However, in spite of such a sea change, the only thing that has retained its reputation is the 'C' language. Even today, millions and billions of students, hobbyists, and professional programmers enjoy the sturdiness, reliability, and user-friendliness of the C language. Today, 'C' enjoys the undisputable recognition in the computing paradigm for diversified applications right from the basic programming, microcontrollers, and spreadsheets to the system programming. Another movement witnessed by the computing paradigm is the Free/Open Source Software (FOSS). The GCC (GNU compiler collection) is extensively regarded as the most imperative part of the FOSS. The GCC holds a full-featured ANSI (American National Standards Institute) C compiler. This compiler provides manifold levels of source code error inspection, thereby helping in extensive debugging. This also leads to optimization on the resulting object code.

In view of the popularity of the 'C', there is no wonder that there are good numbers of books by different publishing houses. However, in my opinion, there is hardly any book focusing on 'C' programming on an open source platform. The proposed book is set to fill this void.

In nutshell, I would like to recommend this book to the young learners owing to its many novel features. Though the market is flooded with a number of good books on 'C' programming, most of them are predominantly focused exclusively on the ANSI 'C' standard. These books are traditional in nature, that is, they start with the routine theory, features of C, data types, Boolean operators, arrays and pointers, etc. In the present book, all such routine theoretical features are skipped. Rather than introducing the underpinning theory, authors have adopted a unique approach of "learning through doing" which I hope will appeal all the young entrants in this promising domain. The 'C' codes given are well supported by easy-to-understand comments wherever required. Mastering the basic modules and hands-on working with the code will enable the reader to grasp the basic building blocks of programs and will definitely lead to build more complex programming tasks. I wish all the potential readers a very happy reading time through this book.

Dr. Arun Patil
Associate Professor of Engineering Management and Education,
Deakin University,
Australia

Preface

The altruistic revolution evidenced by the Computer Science arena is the Free Open Source Software (FOSS) movement, which is now encompassing the entire scholastics world never than ever before. The free licensing and more than that the openness have attracted many software professionals, and there is seemingly an absolute migration of the computer science fraternity from its proprietary counterpart.

However, in spite of the benefits paved by the FOSS such as security, stability, reliability, stability, and cost effectiveness, there are not many screenshotting sources for opening oneself as a programmer in this exciting part of the realm. Undoubtedly 'C' is the widely accepted programming paradigm, and there should have been many books on programming aspects on open source platform, unfortunately which is not the actuality. In view of this, the proposed book is based on open source C programming. C programming is still one of the most prominently used languages even in the era of C++, Java, C#, etc. As C can be used for application level, hardware level, embedded level, graphical level, etc., most of the MNCs use C language to check programming logic of the candidates till today.

Some of the unique features of the proposed book are as follows:

- As per our knowledge, there is not a single book dedicated exclusively for C programming skills using open source compiler, i.e., gcc. Most of the books have C programming as one of the chapter, but this proposed book is fully dedicated to C programming using vi editor and gcc compiler.
- Most of the UG/PG students are not comfortable with Linux platform, but as per the curriculum, C programming is an important component. This proposed book will inculcate and generate interest amongst such students. It will be proved to be a good textbook as well as reference book.
- This proposed book will start with basic gcc environment and programming skills. How to save a C program, how to compile it, and how to execute the complied program will be included in detail. In short, book will cover the programming aspect as self-tutor.

- All the C programming concepts will be presented in the form of programs rather than theory so that students can understand it easily as what they learn they can do it simultaneously.
- The proposed book will showcase actual screenshots of the programs from the programming environment to make it more student-friendly. Because of the user-friendly interface provided in the book, a novice learner can also learn C programming without any difficulty.
- As the book is focused on open source, programs written in this book can execute on different versions of Linux such as Fedora, Ubuntu, and Zorin and can also execute on Turbo C under Windows environment with slight modifications.

In contrast to so many books available on 'C' programming in market, the proposed book harps on hands-on aspects. The experienced faculty members who have devoted themselves for programming in 'C' for more than a decade share their expertise. Some more benefits of the book are as follows:

- Bare minimum theoretical details
- Emphasis on learning by doing
- Coverage right from basics, i.e., data types to advanced concepts such as data structures
- Screenshots of actual output
- Complete program listing which will serve as a software library for future applications
- All the programs based on 'gcc' compiler which is free open source software
- Stimulation of program-oriented learning
- Handy size.

Thus, the proposed book will be useful not only from Computer Science but also from other allied disciplines such as Mathematics, Statistics, Electronics, Physics, Biotechnology, Business Administration, and so on.

K. S. Oza
S. R. Patil
R. K. Kamat

List of Figures

List of Tables

1

C Programming: Logical Continuum of Program of Programming

The very first chapter of this book aims at rolling the notion of programming in 'C'. The 'C' language has evidenced many variations in its life cycle. Considered as the mother of today's programming paradigm, 'C' now exists in variety of its variants even in the structural programming domain such as Handel C, System C, and Impulse C. The rise of 'C' has been phenomenal and led to good number of innovative processes such as simulating the simulations and program the programming. The logical continuum of problem solving in C in terms of systematic progression of flowchart, pseudocode, and structural program gave rise to a generation of passionate programmers. The aftermath of the program, apart from the intended results, and the ease of debugging made 'C' the ultimate choice of the programming community. This chapter unfurls all these details and serves as the starting point of liking the 'C'.

1.1 Linux Operating System and C Language

The Linux operating system and 'C' have an interwoven association. Linux is a well-known open source operating system developed by Finnish undergraduate student Linus Torvalds. Since the release of its first version 0.01 way back in September 1991, it has subsequently became popular in terms of hardware technology and development of software programs. Being a generic operating system, it can be installed on diverse types of hardware produced by different manufacturers. Moreover, it is portable since around 95% part is programmed using C language and this virtue has due to the implicit portability of C which can be harnessed in a high-level, machine-independent manner. Linux supports multiuser programming. It allows us to share data and programs among many users because it takes advantage of available hardware and runs on a lot of diverse chipsets. It has a rich set of utility programs and tools to connect and use these utilities to put together interesting range of applications. The GNU compiler collection (GCC) is a compiler system created under the

GNU Project underneath an assortment of programming languages are being supported. GCC is being distributed by the free software foundation under the GNU General Public License (GNU GPL) and has played a pivotal role in furthering the free software. This main functionality comes due to the fact that though C is a procedure-oriented programming language, it is also a good system language. One can develop compiler as well as operating system using C language. It is a highly structured language. It allows programmers to work with bits and bytes and manipulate them easily. Modularity is yet another key aspect of this language.

This book intends to nurture the development of C programs in Linux Editor to get work done efficiently. Writing C programs in Linux editor paves better performance due to the combined features from C language and Linux operating system. Therefore, it is worthwhile to have a glance at the basics of this programming paradigm.

1.2 Introduction to GCC Compiler

To run the C program under Linux environment, the most commonly compiler required to be used is GNU GCC (www.gnu.org/software/gcc/gcc.html). The GCC compiler is GNU project having C and C++ compiler.

The synopsis and description of GCC compiler can be obtained by using the following command on Linux terminal.

```
************************************************************
# mangcc
Example:
[root@localhostcsd]# mangcc
************************************************************
```

Getting to know the GCC compiler version of your system
In order to get familiar with the version of the C compiler, one can use the following command or (gcc=/usr/bin/gcc).
```
************************************************************
#gcc –version
```
The following commend gives more details regarding the version.
```
#gcc –v
```
Finding the physical location of the C compiler in your system is a weird task, particularly when the system grows in terms of directories, subdirectories, and storage. One can use the following command to get to know the location of the GCC compiler.

#whereisgcc
To know about GCC, use the following command:
#whatisgcc
****

The version utility ensures that the GCC compiler is installed on your system. Though

1.3 Components of Compilation Process

The execution process of C program using GCC compiler is comprised of the following four steps.

Step 1-C Preprocessor, GCC utility calls the C preprocessor
C preprocessor – expands macro definitions and includes header files

Step 2-C compiler, GCC utility calls the C compiler
C compiler – generates assembly language code of the source .c file

Step 3-Assembler, GCC utility calls the assembler
Assembler – generates machine readable code, that is, it converts .c file to .o file, keeping name same as source code file.

Step 4-Linker, GCC utility calls linker
Linker – It links the object modules in the libraries with the program's object modules for which functions are defined in the program. C compiler links the standard C library libc.so (/lib) to handle input and output and general-purpose capabilities.

1.4 Getting Used to the Data Types

Table 1.1 depicts the data types that are used in the C programming.

1.5 Built-In Standard Library

A C header file consists of C function declarations and macro definitions. They have .h extension and can be used between different C source code files. The C preprocessor directive #include is used to include the header file into C source code file.
Example: #include<stdio.h>
Table 1.2 gives the list of header files along with brief description.

Table 1.1   Data types in C

Data Type	Size in Bytes
char	1
short	2
int	4
long	8
ptr	8
long-long	8
float	4
double	8

Table 1.2   Details of header files

File Name	Description
Stdio.h	Defines input/output functions
Stdlib.h	Defines numeric conversion functions, memory allocation functions, etc.
Math.h	Defines mathematical function
String.h	Defines string handling functions
Conio.h	Defines console input/output functions

1.6 Nitty-Gritty of Programming Structures

C programming consists of two basic parts: algorithm and procedure.

1.6.1 Algorithm

Algorithm defines the execution flow of a program. It is a stepwise sequence of instructions which gives solution to the problem. The logical assortment typically found is as follows:

1. Sequence
2. Selection
3. Iteration
4. Case-type statement

In C, "sequence statements" are vital ones, while the decision making is implemented through the "if then else" statement and case-type statements such as "switch statement." The iteration statements are "while," "do-while," and "for" statements.

1.6.2 Pseudocode

Pseudocode is an informal way of writing algorithms. It uses English like language to represent steps in an algorithm. Before any novice programmer starts coding an algorithm, it is always recommended that he/she should write the pseudocode for the same first as it gives good understanding of the algorithm flow. A simple example of the pseudocode is given as below:

```
*************************************************************
If patient temperature is greater than 98 degree
        Print "high fever"
else
        Print "no fever"
*************************************************************
```

1.6.3 Procedure

Procedure is a block of code statements that explain the given algorithm. Procedure consists of statements enclosed within a pair of {}. Some procedures return value to the program construct known as functions.

1.6.4 Program

Program is a set of explicit instructions expressed in terms of programming language. Once we have program as a solution to the problem, then we need to provide some input data to the program to get output.

1.7 C Program Structure

A conceptual view of the structure of C program is as shown below.

This book exclusively uses Vim editor that necessitates the details of the same in nutshell.

1.8 Vim Editor

Vim is Vi Improved, a text editor compatible with Vi. It is common and most natural choice for C programming under Linux environment. We will not go into details of Vim editor here. We are focusing only those modes of Vim editor that are frequently used and most useful for C programming only. One can use the following command on terminal to know more about Vi editor.

```
*************************************************************
#man vi
*************************************************************
```

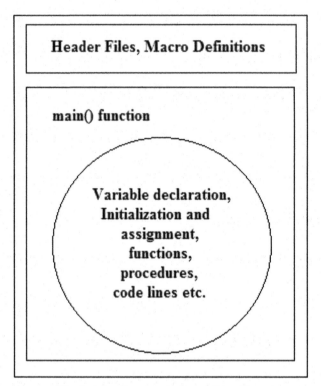

Figure 1.1   Conceptual view of C program.

Figure 1.1 shows conceptual view of a program. It shows the outlook of a C program into two parts part-I will have all global declarations like including header files and preprocessor directives. Part-II is main function which is heart of any C program. It contains all variable declarations, operations on variables, processing statements etc.

Table 1.3 shows different modes and its description.

1.9 Whittling the First 'C' Program

The following steps will lead you to build the first successful 'C' program:

1.9.1 Checking Whether the Compiler Is Working

To check whether your compiler is working or not, run a simple code of lines on terminal as follows.

Table 1.3 Different modes of Vi editor

Mode	Key to be Used	Description
Command mode	—	It is also known as display mode. When you enter into the Vim editor, you are in command mode by default.
Insert mode	i or insert key	This mode is used to enter characters or to write program in Vim editor. Press Esc key on the keyboard to get out from insert mode and enter into the command mode.
Exit mode	a) Esc:wq b) Esc:w c) Esc:q	a) Press esc key followed by :wq to write/save work done and quit from Vim editor and return back to the terminal window. b) Press esc key followed by w to write/save work in the editor and you stay in the same editor. c) Press esc key followed by q to exit from Vim editor without writing or saving any work done and return back to the terminal window.

Program 1.1: First C program to print hello on the screen

At the terminal window, type the following command to open vi editor

[root@localhostcsd]# vi Hello.c   // This will create hello.c C source code file

```
#include<stdio.h>
int main()
{
        printf("\n Hello!Have a good day!\n\n");
        return 0;
}
```

//Press **:wq** to save work done and quit from Vim editor. Now you are on terminal window again.

Compiling Your Program: Type the following command for compilation

[root@localhostcsd]# gccHello.c –o Hello.o

This will create an object file named as Hello.o.

Here, –o parameter is used to create an object file. Using the object file only, we can run the program.

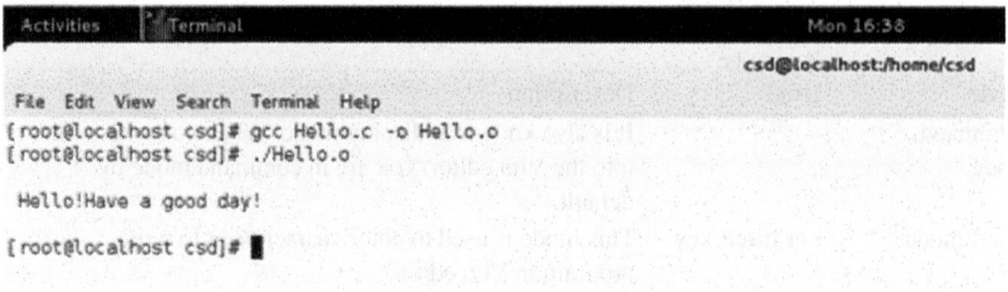

Figure 1.2   Output screen of Hello.c program.

Running Your Program

[root@localhostcsd]# ./Hello.o

This will show the output of the source program.

If it results in the successful execution, then it indicates that your C compiler is working.

As shown in Figure 1.2 the program is compiled and executed successfully with the message Hello! have a good day! with the prompt waiting for next command.

1.10 Execution of Makefile

We can run the C program using Makefile utility also. The following program shows the use of Make utility.

Program 1.2: Stepwise execution of the program using Makefile utility

```
//Enter following command on terminal window and you
will enter in vi editor.

#vi hello.c            //hello.c is the C source file
#include<stdio.h>      //Standard Input /Output files
int main()             //main() always returns int
                          value
{
      printf(''\n Hello!Welcome dear!\n'');
      return 0;
}
```

```
//Press :wq to save work done and quit from vi
   editor. Now you are on terminal window again.

#vi Makefile    //It creates Makefile consisting
                    hello.c source code file

 Hello:hello.c
 gcc -o Hello hello.c
//Press :wq to save work done and quit from vi
   editor. Now you are on terminal window again.

#make           //This will compile Makefile program
 gcc -o Hello hello.c
#./Hello         //This will run Makefile program
Hello!Welcome dear!
```

Makefile execution: Makefile execution consists of creating C source files and executing them with makefile command. Makefile help by creating output file once. It does not need to create the output file more than once if there is no change in the program source code. We can execute the source code directly. It saves memory space as well as compilation time in the execution of C source file.

1.11 Variable Declaration

Variable is a named memory location. Here, variable declaration is same as that of Turbo C compiler.

```
**************************************************************
Syntax: data-type variable-name;
            int number;
            charch;
**************************************************************
```

1.12 Input/Output Statement

The **scanf()** function is used to receive the input from the keyboard. The & variable in scanf() is must and it is called as "Address of" operator. It gives the address of variable in the memory.

The **printf()** function is used to output the values to the screen. We use format specifiers to format the string argument passed as a sequence to the printf() and scanf() functions.

1.13 Format Specifiers

Format specifiers are used to give some specific format or way to accept or print data. They are used with scanf and printf commands. Like if user is entering integer data then scanf should be told about the same by using %d format. Table 1.4 gives details of format specifiers.

1.14 Escape Sequences

When we want to give a different meaning to character or a special character in a C program then that character is preceded by back slash and forms a escape sequence. All the escape sequences starts with back slash(\) as shown in Table 1.5.

Program 1.3: Program to add two integer numbers

```
//Program to demonstrate variable declaration

#include<stdio.h>
int main()
{
     int a,b,c;
     printf(''\n Enter two numbers\n'');
     scanf(''%d%d, &a, &b);
     c=a+b;
     printf(''\n Addition of a and b =%d\n'',c);
     return 0;
}
```

The above program uses three variables, in which the first two accept the numbers to be added and the third one stores the value after addition. Figure 1.3 shows the output of program 1.3.

Table 1.4 Details of format specifiers

Format Specifiers	Characters Matched
%d, %i	Signed integer
%c	Single char
%f, %e	float, exponential
%u	Unsigned integer
%o	Unsigned octal integer
%p	Pointer address (Void*)
%s	Any sequence of non-whitespace characters
%x	Unsigned hex
%ld	Long integer, double

Table 1.5 Showing list of escape sequences

Escape Sequence	Character
\a	Speaker beeps
\b	Backspace
\f	Form feed
\n	New line
\r	Carriage return
\t	Tab
\v	Vertical tab
\\	Backslash
\?	Question mark
\'	Single quote
\"	Double quote

Figure 1.3 Screen showing the output after the execution of the program which is used to add two numbers.

Program 1.4: Program to interchange the values of two variables

```c
/* Swapping two numbers without using third
variable */

#include<stdio.h>
int main()
{
        int first,second;
        printf("\n Enter two numbers\n");
        scanf("%d\n%d",&first,&second);
        printf("\n Before Swapping:%d\t%d\n",
        first,second);
        first=first+second;
        second=first-second;
        first=first-second;
        printf("\n Swapped Numbers: %d\t%d\n\n",
        first,second);
        return 0;
}
```

The above program swaps the values of two variables without using the third variable to store intermediate results. It is saved as 22.c file. Figure 1.4 shows the output of program 1.4 after execution.

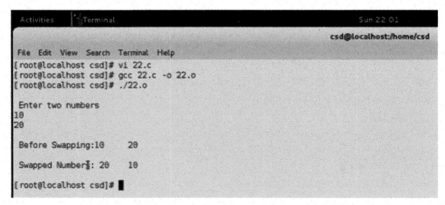

Figure 1.4　Output of the swap program.

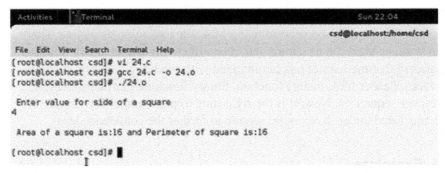

Figure 1.5 Screen showing the output after the execution of the program which is used to calculate area and perimeter of a square.

Program 1.5: The following program accepts the side of a square and calculates its area and perimeter

```
/* Program to calculate area and perimeter of a
square */
#include<stdio.h>
    int main()
{
        int side,area,per;
        printf("\n Enter value for side of a square
        \n");
        scanf("%d",&side);
        area=side*side;
        per=4*side;
        printf("\n Area of a square is:%d and
        Perimeter of squarem is:%d\n\n",area,per);
        return 0;
}
```

Here user is prompted to enter sides of the square using which area is calculated and displayed as shown in Figure 1.5.

1.15 Conclusion

This chapter has introduced the Linux operating system from the viewpoint of C language. The nuts and bolts of both the duo have been put in place for the benefit of the budding programmers. Building a C program under

Linux paradigm has been showcased in a stepwise manner using the GCC compiler. The compilation process using Makefile execution as well as Vim editor without Makefile execution has been effectively depicted. By now, it is expected that the learner has familiarized with the basic concepts such as data types, header files, main() function, library functions, format specifiers, and escape sequences. Now, it is the right time to put together few simple C programs listed under the exercise section to further the confidence level.

1.16 Exercises

1. How to compile a C program using the GCC compiler?
2. Explain the different modes of Vim editor.
3. Write a program to print your biodata.
4. Execute the above program using makefile.
5. Write a program to print the area of triangle.

2

Decision Making and Looping Constructs

2.1 Introduction

Computer programs are executed sequentially, but we can change its flow of execution to bifurcate, repeat code, or take decision. This requires a control mechanism. C language enforces control mechanism by performing set of operations depending upon some condition. All the decisions are dependent on some condition being met. Conditions may be some sort of comparisons to be made to get the right decision. This type of decision making involves the use of relational operators. These operators give the binary result in the form of true and false after comparing values of the variables or constants. Any inbuilt data-type values can be compared using these operators.

Some relational operators are as follows:

- Equal to denoted as ==
- Not equal to denoted as !=
- Less than denoted as <
- Less than or equal to denoted as <=
- Greater than denoted as >
- Greater than or equal to denoted as >=

Programmer has to specify the condition which should be evaluated to as true or false to make a decision. He/she also has to specify the statements to be executed if the condition is evaluated as true and if the condition is evaluated as false. One can give more than one statement to be executed if the condition is evaluated as true or false by enclosing the statements in the curly braces.

C provides two types of flow control statements:

- Branching
- Looping

Depending on evaluating condition, an action is taken and it is called branching.

The number of times certain actions have to be taken is decided by looping.

Branching: Once the condition is evaluated, program flow follows either of the branch, one evaluating to true and second evaluating to false by executing respective statements.

C language has three branching statements: if statement, if-else statement, and switch statement. These statements control the behavior of a program. Each branch consists of set of instructions to be executed depending upon some condition. If that condition evaluates to true, then the corresponding block of statements get executed. The "if statement" is the simplest form. It takes an expression in parenthesis, and if the expression evaluates to true, then the statement/statements get executed; otherwise, they are skipped.

2.2 The if Statement

The if statement is used to make one time decision. Here either the statement is true or false, like binary decision.

General form:

if (condition is true)
       execute statement;

where if is a keyword:

   condition—expression using C's relational operator

   statement—block of code to be executed.

The block of code is executed if and only if the condition is true; otherwise, compiler skips it.

2.3 The if-else Statement

The **if** statement facilitates execution of block of code if condition is true; otherwise, it does nothing. The **if-else** statement provides a way to execute a set of instructions if condition is true or false.

General form:

if (condition is true)
      execute statement;
else
      execute statement;   // if condition is false

Here, **else** part will execute if condition will evaluate to false.

Program 2.1: Use of if-else statement to predict leap year

Leap year comes once in four years. It has 29 days in February. User is asked to enter the year, which is then divided by 4; and if the remainder is zero, then the year is leap year, else not.

```c
/* Program to check the entered year is leap year
or not */

#include<stdio.h>
int main()
{
    int year;
    printf("\n Enter the year\n");
    scanf("%d",&year);
    if(year%4==0)
    {
        printf("\n The entered year is leap year
        \n\n");
    }
    else
    {
        printf("\n The entered year is not leap
        year \n\n");
    }
    return 0;
}
```

It can be observed in Figure 2.1 that the user has entered the year 2013 to check whether it is a leap year, and then, the message "The entered year is not leap year" is shown, as 2013 is not a leap year.

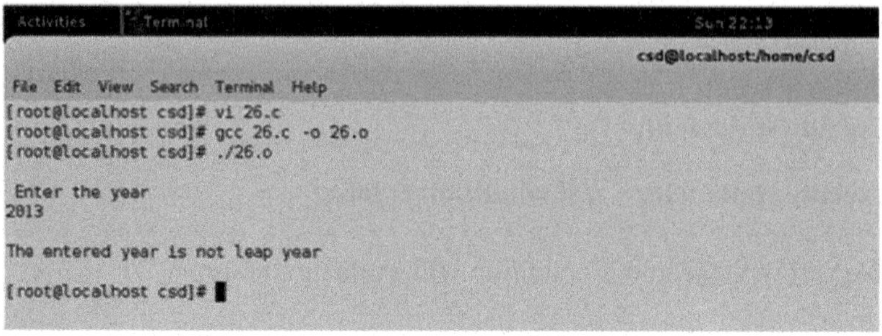

Figure 2.1   Output of leap year program.

2.4 Nested if-else

We can write **if-else** construct inside another **if** or **else** part. This is called as **nesting of if statement.**

General form:

if(condition is true)
        execute statement;
else
{
        if(condition is true)
                execute statement;
        else
                {
                        execute statements;
                }
}

Program 2.2: Use of nested if-else statement to compare two numbers

Here, the user is asked to enter two numbers, and then, the numbers are compared.

```
/* Nested if..else statement program */

#include<stdio.h>
int main()
```

```
{
        int A1,A2;
        printf("\n Enter two integer numbers\n");
        scanf("%d%d",&A1,&A2);
        printf("First number =%d and Second number=
        %d",A1,A2);
        if(A1< A2)
                printf("\n First number is less than
                Second number \n\n");
        else if(A1> A2)
                printf("\n First number is greater
                than Second Number\n\n");
        else
                printf("\n Both are equal\n\n");
        return 0;
}
```

The numbers entered for comparison are 10 and 20, and the output "First number is less than Second number" is displayed. It can be observed in Figure 2.2.

Program 2.3: Use of nested if-else statements

Here, user is asked to enter any character which may be a digit or alphabet or special symbol. This program uses the concept of ASCII value to predict the type.

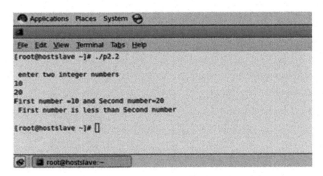

Figure 2.2   Output screen for Program 2.2.

```
/* Program to check whether entered character is a
digit or an alphabet or a symbol */

#include<stdio.h>
int main()
{
        char Alpha;
        printf("\n Enter a character \n");
        scanf("%c",&Alpha);
        if((Alpha>='a' && Alpha<='z')||(Alpha>='A'&&
        Alpha<='Z'))
                printf("\n %c is an alphabate\n\n",
                Alpha);
        else if(Alpha>='0' && Alpha<='9')
                printf("\n %c is a digit\n\n", Alpha);
        else
                printf("\n %c is a symbol\n\n",Alpha);
        return 0;
}
```

It can be observed in Figure 2.3 that when @ is entered, it prints "@ is a symbol", and when 7 is entered, it displays "7 is a digit".

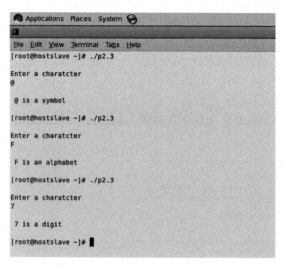

Figure 2.3   Output of Program 2.3.

Program 2.4: Use of if-else statements

Here, user is asked to enter the names of two players and their scores. Using if-else statement, it outputs the name of the player with higher score along with how higher is his score compared to other player.

```c
/* Program to display highest score among two
persons */

#include<stdio.h>
int main()
{
        float sc1,sc2,z;
        char nm1[10],nm2[10];
        printf("\n Enter the first person\n");
        scanf("%s",&nm1);
        printf("\n Enter the score of first
        person\n");
        scanf("%f",&sc1);
        printf("\n Enter the second person \n");
    scanf("%s",&nm2);
        printf("\n Enter the score of second person
        \n");
    scanf("%f",&sc2);
        if(sc1>sc2)
        {
            z=sc1-sc2;
            printf("\n The highest score person is
            %s and he won by the %.2f\n\n",nm1,z);
        }
        else
        {
            z=sc2-sc1;
            printf("\n The highest score person is
            %s and he won by the %.2f\n\n",nm2,z);
        }
        return 0;

}
```

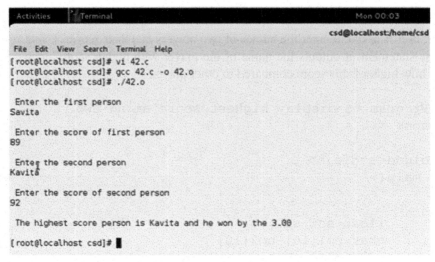

Figure 2.4 Output of Program 2.4 after successful execution.

Figure 2.4 shows the output of program 2.4 after execution.

The names entered by the user are Savita and Kavita with scores 89 and 92, respectively. Then, the output shown is "The highest score person is Kavita and he won by the 3.00" (difference between both the scores).

Program 2.5: Use of if-else statements

Here, user is asked to enter the details of driver such as marital status, gender, and age to predict whether driver is insured or not. For male drivers, if the age is greater than 30 and he is not married, then he is insured; and in case of female drivers, if the age is greater than 25 and she is not married, then she is insured.

```
/* Program to check insured status of driver */

#include<stdio.h>
int main()
{

        char sex,ms;
        int age;
```

```
printf("\n Enter marital status(Y/N),
sex(M/F) and age\n");
scanf("%c%c%d",&ms,&sex,&age);
if(ms=='N'||((sex=='M' && age>30)||(sex=='F'
&& age>25)))
{
        printf("\n Driver is insured\n\n");
}
else
{
        printf("\n Driver is not insured\n
        \n");
}
return 0;

}
```

As can be observed in Figure 2.5, user has entered marital status as Y (i.e., the driver is married), gender as F (i.e., female), and age as 25. Then, the output "Driver is not insured" is shown.

2.5 The Switch Case Statement

The switch case statement is a case–control statement where decision is made in number of choices. So it is called as **switch**.

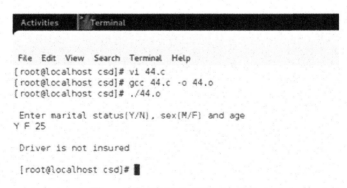

Figure 2.5 Output of Program 2.5.

General form:

switch(expression)

{

      case const-1:

              statement/s;
              break;

      case const-2:

              statement/s;
              break;

      case const-n:

              statement/s;
              break;

      default:

              statement/s;

}

where **expression** is any expression that evaluates to integer value.

When there is a match between **case constant** value and **expression** value, the corresponding statement gets executed and **break** statement takes program control out of the **switch**.

Program 2.6: Use of switch case statement

Use of switch statement to print day of the week. User is prompted to enter a number between 1 and 8. If 1 is entered, Sunday is displayed, and if 2 is entered, Monday is displayed, and so on. If user enters 8, the program execution will be terminated.

```c
/* Program showing use of switch case statement */
#include<stdio.h>
int main()
{
        int ch;
        printf("\n Enter Your Day...\n");
```

```
printf("\n1.Sun\n2.Mon\n3.Tues\n4.Wed\n5.
Thurs\n6.Fri\n7.Sat\n8.Exit\n\n");
scanf("%d",&ch);
switch(ch)
{

        case 1:printf("\n You have selected
        Sunday\n\n");
        break;
        case 2:printf("\n You have selected
        Monday\n\n");
        break;
        case 3:printf("\n You have selected
        Tuesday\n\n");
        break;
        case 4:printf("\n You have selected
        Wednesday\n\n");
        break;
        case 5:printf("\n You have selected
        Thursday\n\n");
        break;
        case 6:printf("\n You have selected
        Friday\n\n");
        break;
        case 7:printf("\n You have selected
        Saturday\n\n");
        break;
        case 8:return 0;
        default:printf("\n You have selected
        wrong day...\n\n");

}
        return 0;

}
```

It can be observed in Figure 2.6 that a menu is displayed with the number and the corresponding day of the week. When 5 is entered, the output "You have selected Thursday" is shown.

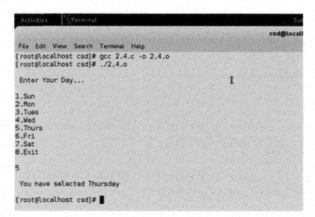

Figure 2.6   Output screen after execution of Program 2.6.

Program 2.7: Use of switch case statement to convert seconds to hours and minutes and vice versa

It is a menu-driven program where menu is displayed on the screen asking user to select the menu. There are four menu options for inter-conversion of time units, and last option is to exit from the program. Switch case statement is used to interpret user's choice, and accordingly, the conversion takes place.

```c
/* Program for time conversion */

#include<stdio.h>
int main()
{
        float hr,sec,min,time;
        int ch;
        do
        {
        printf("\n Enter your choice=\n 1.Second in
        to Hour\n 2.Second in to minute\n 3.Hour in
        to Second\n 4. Minute in to Second\n
        5. Exit\n");
        scanf("%d",&ch);
        switch(ch)
        {
```

```c
case 1:
     printf("\n Enter time in
     second");
     scanf("%f",&time);
     hr=time/3600;
     printf("\n Time in Hour:%f\n
     \n",hr);
     break;
case 2:
     printf("\n Enter time in
     Second");
     scanf("%f",&time);
     min=(time/3600)*60;
     printf("\n Time
     in Minute is:%f\n\n",min);
     break;
case 3:
     printf("\n Enter time in
     Hour");
     scanf("%f",&time);
     sec=time*3600;
     printf("\n Time in Second:
     %f\n\n",sec);
     break;
case 4:
     printf("\n Enter time in
     Minute");
     scanf("%f",&time);
     sec=time*60;
     printf("\n Time in Second:
     %f\n\n",sec);
     break;
case 5:
     return 0;
     break;
default:
     printf("\n Enter Valid
     Choice");
     break;
```

```
        }
        }while(1);
        return 0;
}
```

When user executes the program, a menu is displayed as shown in Figure 2.7 asking user to enter his choice. User has entered 2 indicating conversion of seconds to minutes. Then, user is prompted to enter time in seconds. It is entered as 120, and after conversion, the result is displayed as 2 minutes. User exits from the program by entering 5 as his choice.

Program 2.8: Another example showing the use of switch case statement to implement simple arithmetic calculator

Here, user is asked to enter two values. Then, a menu is displayed showing possible operations on those two values. User is asked to select the menu, and accordingly, the selected operation is carried out on those two values and the result is displayed.

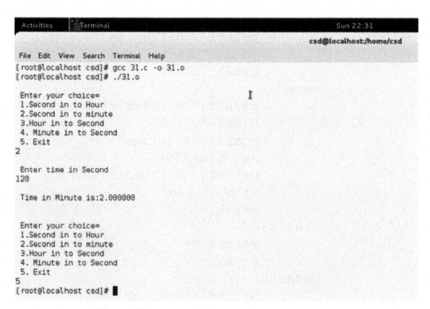

Figure 2.7   Output after execution of Program 2.7.

```
/* Program to implement simple calculator using
switch case statement */

#include<stdio.h>
int main()
{
        float n1,n2,result
        int p,d,r,i;
        char s;
        printf("\n Enter the two values:\n");
        scanf("%f%f",&n1,&n2);
        do
        {
                printf("\n Select any Arithmetic
                Operation: \n 1:+ \n 2:- \n 3:* \n
                4:/ \n 5:% \n 6:Exit\n");
                scanf("%d",&i);
                switch(i)
                {
                        case 1:
                                result=n1+n2;
                                printf("\n Addition
                                is:%.2f\n\n",result);
                                break;
                        case 2:
                                result=n1-n2;
                                printf("\n Subtraction
                                is:%.2f\n\n", result);
                                break;
                        case 3:
                                result=n1*n2;
                                printf("\n
                                Multiplication is:%.2f
                                \n\n", result);
                                break;
                        case 4:
                                result=n1/n2;
                                printf("\n Division
                                is:%.2f\n\n",result);
```

```
                                break;
                  case 5:
                                printf("\n Enter two
                                integer values:\n");
                                scanf("%d%d",&p,&d);
                                r=p%d;
                                printf("\n Modulo
                                Division is:%d\n\n",r);
                                break;
                  case 6:
                                return 0;
                                break;
                    default:
                                printf("\n Select
                                Valid Operation\n");
                                break;
              }
        }while(1);
        return 0;
}
```

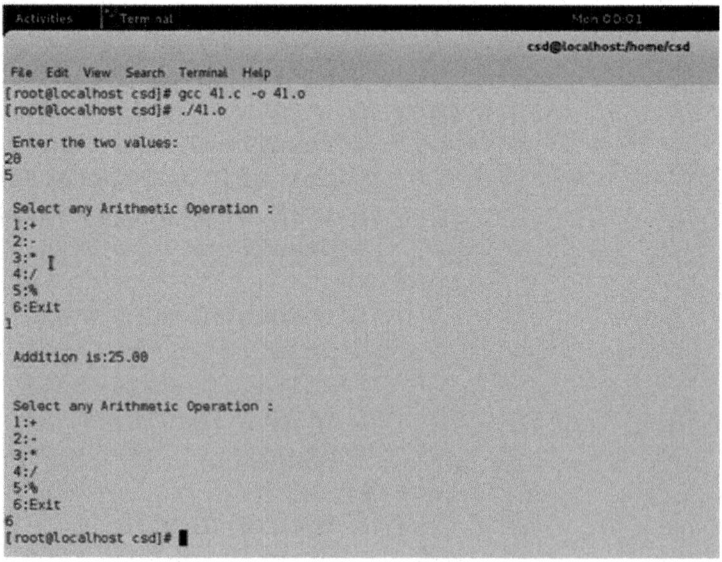

Figure 2.8 Output of calculator program.

As shown in Figure 2.8, user enters the values as 20 and 5, and then, a menu is displayed. Then, user selects menu item 1 indicating + operation to be carried out, and the result of the addition, that is, 25, is displayed. User quits the program by entering 6 as menu option.

Program 2.9: Switch case statement for simple bank transactions

The three important transactions that any bank carry out are deposit, withdrawal, and balance showing. It is a menu-based program asking user whether they want to deposit, withdraw, or check balance. Balance is shown after each transaction.

```c
/* Program to show bank transactions */

#include<stdio.h>
int main()
{
        float bal,d1,w1;
        char ch;
        int i;
        do
        {
            printf("\n Enter your choice:
            1. Deposit amount\n 2.Withdraw amount
            \n 3.Show balance\n 4.Exit");
            scanf("%d",&i);
            switch(i)
            {
                case 1:
                        printf("\n Enter your
                        amount to deposit into
                        balance");
                        scanf("%f",&d1);
                        bal=bal+d1;
                        printf("\n The total
                        balance is:%.2f\n\n",
                        bal);
                        break;
```

```
                        case 2:
                                printf("\n The
                                available balance is:
                                %f", bal);
                                printf("\n Enter
                                withdrawal amount");
                                scanf("%f",&w1);
                                bal=bal-w1;
                                if(bal<0)

                                {

                                 printf("\n Negative
                                 balance is:%.2f\n\n",
                                 0-bal);

                                }
                                else
                                 printf("\n Remaining
                                 balance is:%.2f\n\n",
                                 bal);
                                break;
                        case 3:
                                printf("\n Balance is:
                                %.2f\n\n",bal);
                                break;
                        case 4:
                                return 0;
                                break;
                        default:
                                printf("\n Enter valid
                                option");

                }
        }while(1);
        return 0;


}
```

After execution of the program, a menu is displayed asking user about the type of transaction they want to carry out on their account. As shown in Figure 2.9, user selects 1 to deposit amount in his account. After selecting 1 as menu option, user is prompted to enter the amount to be deposited, and 10000 is entered by the user. Total balance is shown after depositing the amount 10000.

If we want to execute some statement(s) repetitively for fixed number of times, then we should use looping statements.

Figure 2.9 Output screen of Program 2.9.

2.6 The while Loop

As the name suggest it allows to do something repeatedly till the condition specified is true, once the condition is false loop terminates. In short it repeatedly allows to execute set of statements till the given condition is true.

General form:

initialize loop counter;
while(condition is true)   // testing of loop counter using condition
{

       execute statements;
       increment loop counter;

}

Here, condition is tested first, and if it evaluates to true, then body of loop is executed till the condition is valid.

Program 2.10: Use of while statement

Following program checks whether the entered number is an Armstrong number or not. A n-digit number is said to an Armstrong number if sum of nth powers of its digits is equal to the number itself.

   Example number 370 is a three-digit number, so n = 3. Now take cube of all its digits and sum, so you get the same number back. $370 = 3^3 + 7^3 + 0^3$.

   371 is a three-digit number, so n = 3 and it is an Armstrong number $371 = 3^3 + 7^3 + 3 = 27 + 343 + 1$.

   1634 is also an Armstrong number with n = 4. $1634 = 1^4 + 6^4 + 3^4 + 4^4 = 1 + 1296 + 81 + 256$.

```
/*Program to check the entered number is Armstrong
number or not*/

#include<stdio.h>
int main()
{
        int ans,rem,no,temp;
        ans=0
```

```
printf("\n Enter any number\n");
scanf("%d",&no);
temp=no;
while(no>0)
{
        rem=no%10;
        ans=ans+rem*rem*rem;
        no=no/10;
}
if(ans==temp)
{
        printf("\n The entered number is
        Armstrong Number\n\n");
}
else
{
        printf("\n The entered number is not
        Armstrong Number\n\n");
}
return 0;
}
```

Following Figure 2.10 shows the output of program 2.10 after execution.

The number entered by user is 153 which is a three-digit number, so sum of cube of its digits is the number itself and it is an Armstrong number.

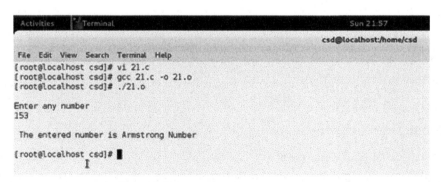

Figure 2.10   Output of Armstrong program.

Program 2.11: Computing factorial of a positive number

Factorial of a number is product of all integers less than or equal to number itself but greater than or equal to 1. It is denoted by exclamation (!) symbol. For example, factorial of 4 is equal to 24: 4! = 4*3*2*1 = 24 1! = 1, 2! = 2*1 = 2, 3! = 3*2*1 = 6 ...

```c
/* Program to display factorial of user entered
number */

#include<stdio.h>
int main()
{
        int temp,Fact,no;
        fact=1;
        printf("\n Enter a positive Number\n");
        scanf("%d",&no);
        temp=no;
        while(temp>0)
        {
                Fact=Fact*temp;
                temp--;
        }
        printf("\n The Factorial of given number is
        =%d\n\n",Fact);
        return 0;
}
```

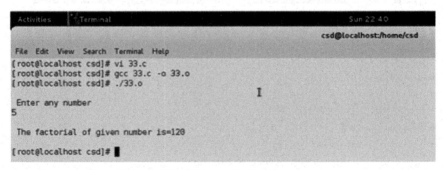

Figure 2.11 Output of factorial program.

In the above output screen of Figure 2.11, if the user enters number 5, its factorial will be 5*4*3*2*1 = 120, which is printed on the screen as factorial of 5.

Program 2.12: Checking whether the number entered by user is prime or not using while loop

A number is said to be prime if it is evenly divided only by itself or one.

For example, 5 is a prime number as it can be evenly divided by 5 itself or 1, whereas 8 is not a prime number as it can be evenly divided by 2 into four parts, that is, 8 = 2+2+2+2; and even into two even parts of four, that is, 8 = 4+4. Some of the prime numbers are 1, 2, 3, 5, 7, and 11.

```c
/* Prime number checker */

#include<stdio.h>
int main()
{

        int pnum,d;
        d=2;
        printf("\n Enter any number\n");
        scanf("%d",&pnum);
        while(d<pnum)
        {
                if(pnum%d==0)
                {
                        Printf("\n The Number is not
                        Prime number\n\n");
                        break;
                }
                i++;
        }
        if(d==pnum)
                printf("\n The Number is Prime
                number\n\n");
        return 0;

}
```

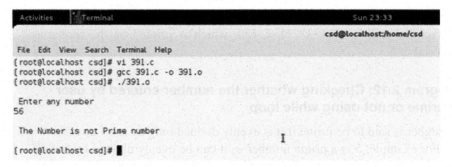

Figure 2.12 Output of prime number checker program.

Figure 2.12 shows the output of prime number checker program.

After execution of the Program 2.12, the user has entered number 56 to check whether it is a prime number or not. The output displayed is "The Number is nor Prime number" as it can be evenly divided into 7 equal parts of 8 each. 56 = 8+8+8+8+8+8+8

Program 2.13: Use of while loop to convert a decimal number to binary form

A number to base 10 is a decimal number with possible values from 0 to 9, and a number to the base 2 is a binary number with the possible values 0 & 1. To convert a decimal number to the binary number, divide the decimal number by 2 and store the remainder which will be either 0 or 1. Continue to divide and store the remainder till the quotient becomes 0. Now all the remainders in reverse order will be the binary equivalent of decimal number.

Example: Binary equivalent of decimal number 12 is = 1100

$12\%2 = 0$
$6\%2 = 0$
$3\%2 = 1$
$1\%2 = 1$

```
/* Decimal to binary convertor */

#include<stdio.h>
int main()
{
        int Num,i,Remainder,a,Rev;
```

```
a=1;
rev=0;
printf("\n Enter decimal number\n");
scanf("%d",&Num);
while(Num>0)
{
        Remainder =no%2;
        Rev=Rev+Remainder*a;
        Num=Num/2;
        a=a*10;
}
printf("\n The binary conversion is:%d\n\n",
Rev);
return 0;
}
```

The number entered by user is 10 and the its binary output is displayed as 1010 as shown in Figure 2.13. Output can be verified using the above conversion method.

Program 2.14: Using while loop to convert decimal number to octal number

Here in this program, octal number has the possible values between 0 and 7. Conversion process is same as shown in Program 2.13 with a small change. Instead of dividing the number by 2, it is divided by 8 with remainders in the range of 0 to 7.

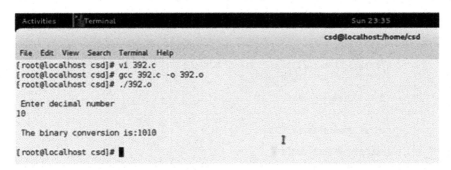

Figure 2.13 Output screen for decimal to binary conversion program.

```
/* Program to convert decimal number to octal
number */

#include<stdio.h>
int main()
{
        int no,i,Remainder,Rev,a;
        Rev=0;
        a=1;
        printf("\n Enter decimal number\n");
        scanf("%d",&no);
        while(no>0)
        {
                Remainder=no%8;
                Rev=Rev+Remainder*a;
                no=no/8;
                a=a*10;
        }
        printf("\n Octal conversion is:%d\n\n",Rev);
        return 0;
}
```

Here, as shown in Figure 2.14, user is asked to enter a decimal number for conversion. User has entered 10 and its octal equivalent is given as 12, which is calculated as shown below.

10%8 = 2
1%8 = 1

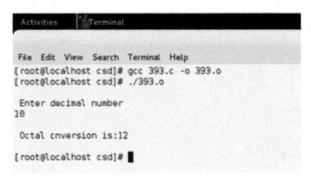

Figure 2.14   Output screen of octal conversion program.

2.7 The odd Loop

When we do not know how many times, we have to execute the statements, and then, we should use odd loop. It is called as odd loop because we do not know the number of iterations in advance.

General form:

do
{
    execute statements;
}while(condition);

Here, first statements are executed, and then, condition is tested, and the loop body executed until condition becomes false.

Program 2.15: Use of odd loop to display employee details

Here, first, user is prompted to enter employee details such as name, department, and salary.

    The data entered by user is printed using printf statements. User will be prompted to enter the data till ch variable value is equal to 'y'. As soon as user enters any other value, loop will be terminated.

```c
/* Program to calculate power of a given number */

#include<stdio.h>
int main()
{
        int ans,pow,no,j;
        ans=1;
        printf("\n Enter any number and power\n");
        scanf("%d%d",&no,&pow);
        printf("\n Number:%d and Power=%d\n",no,pow);
        for(i=1;i<=pow;i++)
        {
                ans=ans*no;
        }
        printf("\n Answer is:%d\n\n",ans);
        return 0;

}
```

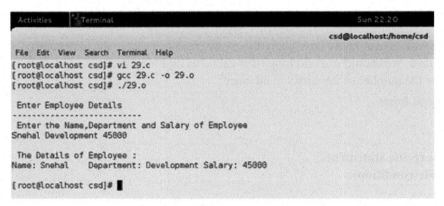

Figure 2.15   Output screen for odd loop.

If one looks at the code of Program 2.15 carefully on execution, he/she can see that the main loop control variable in the program is ch. It is a character variable which is not initialized but only declared. But still when you execute your program, you get the output. This is because odd loop executes at least once before checking the condition. Even if the condition is not satisfied, it will execute once.

As discussed in the above paragraph, the loop executes once asking employee details and printing the entered details as shown in Figure 2.15. But it does not ask for the details second time as the loop is terminated due to no value assigned to the loop control variable ch.

Program 2.16: Use of do while loop to simulate power function

Here, user is asked to enter the two numbers: let's say x and y. Power is y times x, denoted as x^y. In this program, both are positive integers.

```c
/* Program to calculate power of a given number
using do...while loop */

#include<stdio.h>
int main()
{
        int no,pow,res,i;
        res=1;
        i=0;
```

```
printf("\n Enter the number and power\n");
scanf("%d%d",&no,&pow);
do
{

        res=res*no;
        i++;

}
while(i<pow);
printf("\n The power of a given number is:
%d\n\n",res);
return 0;

}
```

Here, as shown in Figure 2.16 the two numbers entered are 2 and 5, then the output becomes 2^5, which is equal to 5 times 2, that is, 32.

Program 2.17: Use of switch case in do. . .while loop

This program is similar to Program 2.8, which is used to simulate a simple calculator in which the user has to enter the two numbers after selecting the operation. But here in this program, user has to enter the two numbers before entering the loop and he can carry out all the arithmetic operations on these two numbers.

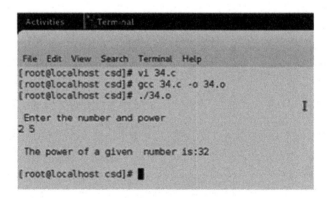

Figure 2.16 Output of power computing program.

```c
/* Program for arithmetic operations using
do...while loop */

#include<stdio.h>
int main()

{
        int n1,n2,choice;
        printf("\n Enter two numbers");
        scanf("%d%d",&n1,&n2);
        do
        {
                printf("\n Enter Your Choice:
                \n 1.Addition \n 2.Subtraction
                \n 3.Multiplication \n 4. Division
                \n 5.Exit");
                scanf("%d",&choice);
                switch(choice)
                {
                        case 1:
                                printf("\n The
                                Addition is:%d\n\
                                n",n1+n2);
                                break;
                        case 2:
                                printf("\n The
                                Subtraction is:
                                %d\n\n", n1-n2);
                                break;
                        case 3:
                                printf("\n The
                                Multiplication is:
                                %d\n\n",n1*n2);
                                break;
                        case 4:
                                printf("\n The
                                Division is:%d\n\n",
                                n1/n2);
                                break;
```

```
case 5:
        return 0;
default:
        printf("\n Enter
        Valid Choice");
    }
}
while(1);
return 0;
}
```

It can be observed in Figure 2.17 that user has entered the two numbers as 2 and 5, and then, the menu is displayed with operations which can be carried out on these two numbers. User selects Multiplication, and the result 10 is displayed. Again, user is asked to select a menu option but not asked to enter the numbers again as in the case of Program 2.8.

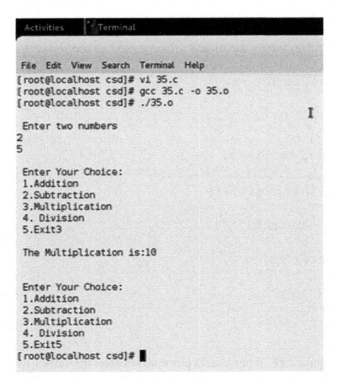

Figure 2.17 Output screen of Program 2.17 after execution.

2.8 The for Loop

The for loop allows counter initialization, test condition, and counter increment in single statement.

General form:

for (initialize counter; test counter; increment counter)
{
    execute statements;
}

Program 2.18: Simulation of power function using for loop

We have already seen simulation of power function using do…while loop in Program 2.16.

Here, do...while loop is replaced by for loop.

```
/* Program to draw pattern */

#include<stdio.h>
int main()
{
    int i,j;
    for(i=1;i<=5;i++)
    {
        for(j=1;j<=i;j++)
        {
                printf("*");
        }
        printf("\n");
    }
    printf("\n\n");
    return 0;
}
```

As shown in Figure 2.18, When user is prompted to enter the values, the power 3 and number 3 are entered. Then, the output is 3^3, which equals 27.

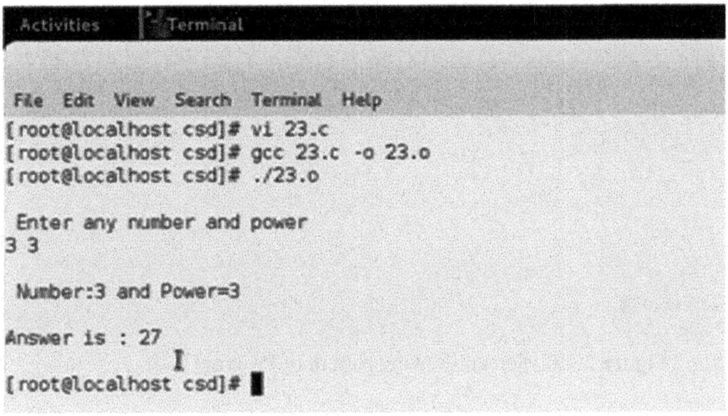

Figure 2.18 Output screen for Program 2.18.

Program 2.19: Use of nested for loop to print a pattern on screen

This program draws a right angle triangle filled with stars.

```
/* Program to draw pattern */
#include<stdio.h>
int main()
{
    int i,j;
    printf("\n");
    for(i=1;i<=5;i++)
    {
        for(j=1;j<=i;j++)
        {
            print("*");
        }
        printf("\n");
    }
    printf("\n\n");
    return 0;
}
```

Here, as shown in Figure 2.19 there is no user interaction, and as soon as the program executes, the pattern is drawn on the screen as output of the program.

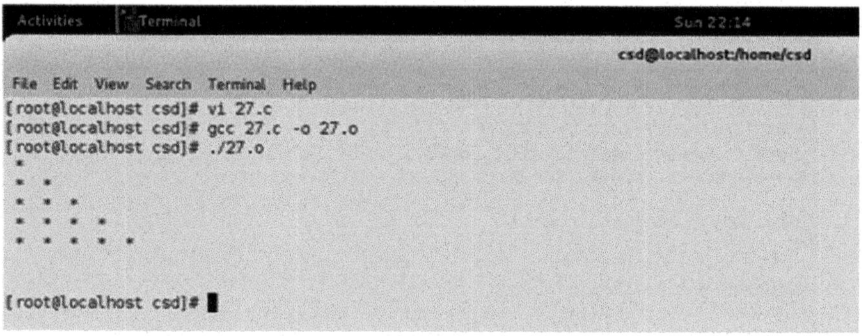

Figure 2.19   Screen showing pattern of Program 2.19.

Program 2.20: Use of nested for loop for another pattern on screen

Here, the program is same as Program 2.19, but instead of creating a right angle triangle with asterisks, it uses numbers that increase by one at each layer.

```c
/* Program to draw pattern */
#include<stdio.h>
int main()
{
        int i,j;
        printf("\n");
        for(i=1;i<=5;i++)
        {
                for(j=1;j<=i;j++)
                {
                        print("%d",j);
                }
                printf("\n");
        }
        printf("\n\n");
        return 0;
}
```

Here, as shown in Figure 2.20 no user interaction is required after execution, and the triangle formed with numbers is displayed.

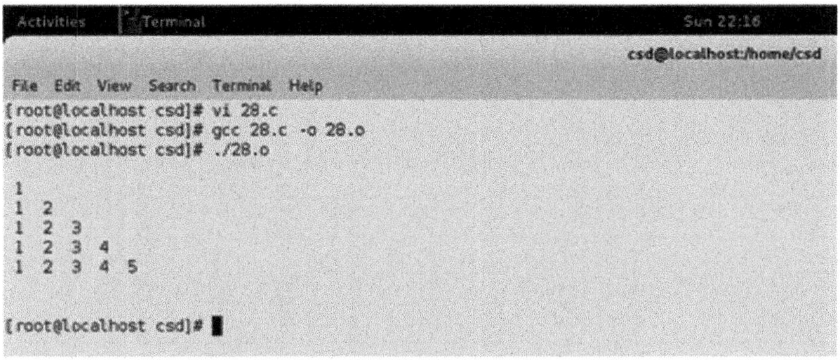

Figure 2.20 Output of pattern program.

Program 2.21: Use of nested do...while and for loop

Here, while is used inside do…while loop to print multiplication tables from 1 to 10. Here, no user interaction is required.

```c
/* Program to display multiplication table from 1
to 10 */

#include<stdio.h>
int main()
{
        int i,res,no;
        no=1;
        printf("\n\n");
        do
        {
                for(i=1;i<=10;i++)
                {
                        res=i*no;
                        printf("\t%d",res);
                }
                printf("\n");
                no++;
        }while(no<=10);
        printf("\n\n");
        return 0;
}
```

```
Activities      Terminal                                          Sun 22:39

                                              csd@localhost:/home/csd

File Edit View Search Terminal Help
[root@localhost csd]# vi 32.c
[root@localhost csd]# gcc 32.c -o 32.o
[root@localhost csd]# ./32.o                     I

          1      2      3      4      5      6      7      8      9     10
          2      4      6      8     10     12     14     16     18     20
          3      6      9     12     15     18     21     24     27     30
          4      8     12     16     20     24     28     32     36     40
          5     10     15     20     25     30     35     40     45     50
          6     12     18     24     30     36     42     48     54     60
          7     14     21     28     35     42     49     56     63     70
          8     16     24     32     40     48     56     64     72     80
          9     18     27     36     45     54     63     72     81     90
         10     20     30     40     50     60     70     80     90    100

[root@localhost csd]#
```

Figure 2.21   Output showing tables from 1 to 10.

Here, as shown in Figure 2.21 it can be observed that for loop actually computes the tables and do while loop is used for formatted printing and incrementing the counter.

Program 2.22: Use of nested if and for loop

This program computes remainder without using mod operator. Multiplication and subtraction operations are used for computing remainder.

The outer if loop first checks whether the dividend is greater than the divisor, and if it is so, then only it will enter for loop or else program will return 0.

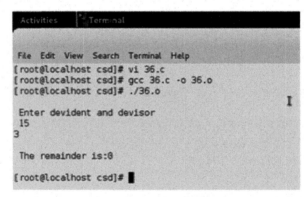

Figure 2.22   Output screen of Program 2.22.

```
/* Program to calculate remainder without using
mod(%) operator */

#include<stdio.h>
int main()
{
        int i,dev,div,a,rem;
        printf("\n Enter dividend and divisor\n");
        scanf("%d%d",&dev,&div);
        if(dev>div)
        {
                for(i=1;dev>=a;i++)
                {
                        a=div*i;
                        if(a<=dev)

                        {
                                rem=dev-a;

                        }
                }
        }
        printf("\n The remainder is:%d\n\n",rem);
        return 0;
}
```

Here, as shown in Figure 2.22 user is prompted to enter dividend and divisor. User enters 15 as dividend and 3 as divisor. The remainder is printed as 0 as 3 evenly divides 15.

Program 2.23: Use of nested if – for statements to divide two numbers without using division operator /

Here, division is carried out using multiplication and subtraction operations. Quotient is printed as output if dividend is greater than divisor.

```
/* Program to calculate division of two numbers
without using division(/) operator */

#include<stdio.h>
int main()
```

```c
{
        int i,dev,div,a,quot,rem;
        printf("\n Enter dividend and divisor\n");
        scanf("%d%d",&dev,&div);
        if(dev>div)
        {
                for(i=1;dev>=a;i++)
                {
                        a=i*div;
                        fi(a<=dev)
                        {
                                rem=dev-a;
                                quot=i;
                        }
                }
        }
        else
                printf("\n The divisor is greater
                than dividend\n");
        printf("\n The quotient is=%d\n\n",quot);
        return 0;
}
```

Here, as shown in Figure 2.23 user enters 20 as dividend and 5 as divisor. Quotient is computed as 4.

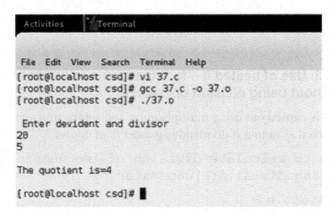

Figure 2.23   Output of Program 2.23.

Program 2.24: Use of two non-nested for loops in the program

Here, the first for loop is used to print the series as: 1+1/2+1/3+..... And, the second for loop is used to compute the series.

```c
/* Program to display sum of series */
#include<stdio.h>
int main()
{
        float ans,i;
        int n,j;
        ans=1.0;
        printf("\n Enter a number, which indicates
        sequence number for addition\n");
        csanf("%d",&n);
        printf("1");
        for(j=2;j<=n;j++)
        {
                printf("+1/%d",j);
        }
        for(i=2.0;i<=n;i++)
        {
                ans=ans+(1/i);
        }
        printf("\n Answer =%f\n\n",ans);
        return 0;
}
```

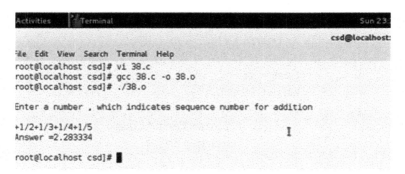

Figure 2.24   Output of Program 2.24.

Here, as shown in Figure 2.24 user is prompted to enter the number of elements in the series. User has entered 5, so series with 5 elements is displayed and computed to print the answer.

Program 2.25: Use of nested for and while loop to sum series

Here, series is the same as computed in Program 2.24 except that factorial is used in this program.

Again, the first for loop is used to print the series on the screen as: 1+ 1/2! + 1/3! + 1/4! .....

The second for loop actually computes the series, and while loop computes the factorial.

```
/* Program to display sum of series */
#include<stdio.h>
int main()
{
        float ans,temp,fact,i;
        int n,j;
        ans=1.0;
        fact=1.0;
        printf("\n Enter any number, indicating
        sequence number\n");
        scanf("%d",&n);
        printf("1");
        for(j=2;j<=n;j++)
        {
                printf("+1/%d",j);
                printf("!");
        }
        for(i=2.0;i<n;i++)
        {
                temp=i;
                while(temp>0)
                {
                        fact=fact*temp;
                        temp--;
                }
                ans=ans+(1/fact);
        }
```

```
        printf("\n Answer=%f\n\n",ans);
        return 0;
}
```

Here, as shown in Figure 2.25 the number of elements in series is entered as 5 and the computed series result is displayed.

Program 2.26: Use of non-nested for loops to compute sum of even and odd numbers in a set

Here some integers are entered by the user and stored in an array by the first for loop.

The second for loop reads the stored numbers and checks whether they are odd or even: if they are odd, they are summed with odd numbers, and similarly, the even numbers are also summed together.

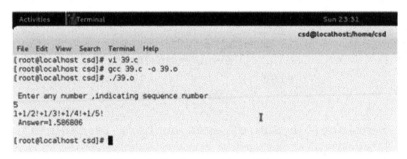

Figure 2.25   Output of factorial series.

Figure 2.26   Output of even and odd sum program.

```
/* Program to calculate sum of even numbers and sum
of odd numbers in a set of values */

#include<stdio.h>
int main()

{
        int Num[5],i,even,odd;
        even=0;
        odd=0;
        printf("\n Enter the elements:\n");
        for(i=0;i<5;i++)
        {
                scanf("%d",&Num[i]);
        }
        for(i=0;i<5;i++)
        {
                if(Num[i]%2==0)
                        even=even+Num[i];
                else
                        odd=odd+Num[i];
        }
        printf("\n Sum of even numbers is:=%d\n\n",
        even);
        printf("\n Sum of odd numbers is:=%d\n\n",
        odd);
        return 0;
}
```

Here, as shown in Figure 2.26 user is prompted to enter 5 numbers, which are then checked for even and odd, and accordingly, the even sum and odd sum are displayed.

Program 2.27: To check whether the number is positive, negative, or zero in a set of numbers

Here, integers are accepted and stored in the array. Then, these integers are read one by one. If the integer is greater than zero, then it is positive; else if the integer is less than zero, then it is negative, or else the integer is zero. Separate counters are maintained for positive, negative, and zero integers.

```c
/* Program to count positive integers, negative
integers and zeros in a given set of integers */

#include<stdio.h>
int main()
{
        int A[10],i,positive,negative,zero;
        positive=0;
        negative=0;
        zero=0;
        printf("\n Enter Array elements\n");
        for(i=0;i<10;i++)
        {
                scanf("\n %d",&A[i]);
        }
        for(i=0;i<10;i++)
        {
                if(A[i]>0)
                        positive=positive+1;
                else if(A[i]<0)
                        negative=negative+1;
                else if(A[i]==0)
                        zero=zero+1;
        }
        printf("\n The positive numbers are:=%d\n
        \n",positive);
        printf("\n The negative numbers are:=%d\n
        \n",negative);
        printf("\n The number of zeros:=%d\n\n",
        zero);
        return 0;

}
```

Here, as shown in Figure 2.27 user is prompted to enter 10 integers. They are checked for positive, negative, and zero, and accordingly, their counters are

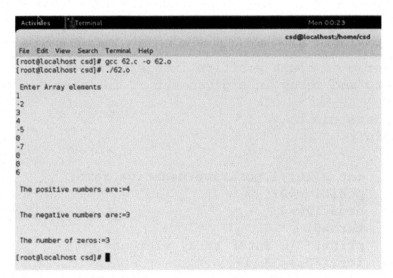

Figure 2.27 Output screen of Program 2.27.

increased. This program prints the total number of positive integers, negative integers, and zeroes.

Program 2.28: To find the maximum number from a set of numbers

Here, a set of numbers are stored in an array. A variable no. is initialized to zero. Now these stored numbers are compared with no., and if the stored number is greater than no., then that number is copied to no. variable. Value of no. at the end is the maximum number from the set.

```
/* Program to find maximum number from a set
of numbers */

#include<stdio.h>
int main()

{
        int i,no=0;
        int a[5];
        printf("\n Enter numbers\n");
        for(i=0;i<5;i++)
```

```
{
        scanf("%d",&a[i]);
}
for(i=0;i<5;i++)

{
        if(a[i]>no)
        {
                no=a[i];
        }

}

printf("\n Maximum number is=%d\n\n",no);
return 0;
}
```

Here, user has entered 5 numbers in the range 10 to 50 as shown in Figure 2.28.
The maximum of all, that is, 50, is displayed as the maximum number.

Program 2.29: To compute square root of a number iteratively

Here, nested for and if statements are used for computing the square root of a
number. It works correctly only for positive integers.

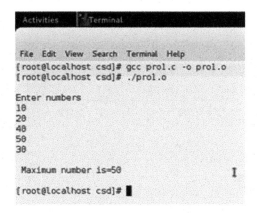

Figure 2.28   Output screen of Program 2.28.

```
/* Program to find square root of a number */

#include<stdio.h>
int main()
{
        int no,i;
        printf("\n Enter number\n");
        scanf("%d",&no);
        for(i=no;i>0;i--)
        {
                if(no%i==0)
                {
                        if(i*i==no)
                        printf("\n The square root of
                        given number=%d\n\n",i);
                }
        }
        return 0;
}
```

User is prompted to enter a number for computing the square root, and the number 25 is entered as shown in Figuer 2.29. The square root of 25 is displayed as 5.

Program 2.30: Use of for loop to add *n* numbers

Here, the first for loop is used to accept *n* numbers and the second for loop to compute their sum.

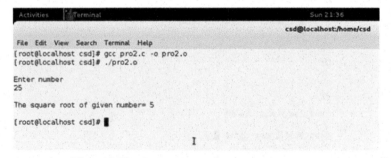

Figure 2.29   Output screen of square root program.

```c
/* Program to display sum of n numbers */

#include<stdio.h>
int main()

{
        int no,sum,num[10];
        sum=0;
        int i;
        printf("\n How many numbers you want to
        enter?\n");
        scanf("%d",&no);
        printf("\n Enter numbers:\n");
        for(i=0;i<no;i++)
        {
                scanf("%d",&num[i]);
        }
        for(i=0;i<no;i++)
        {
                sum=sum+num[i];
        }
        printf("\n The sum of n numbers is=%d\n\n",
        sum);
        return 0;

}
```

Here user is prompted to enter the number of integers to be entered. As seen in Figure 2.30, 5 is entered, and then, the user is prompted to enter 5 integers. Finally, the sum is displayed.

Program 2.31: To display multiplication table of the entered number

Printing of multiplication table from 1 to 10 is already discussed in Program 2.21, but there was no user interface and everything was in code. Here, it is user-based and program will display the multiplication table of any number entered by the user.

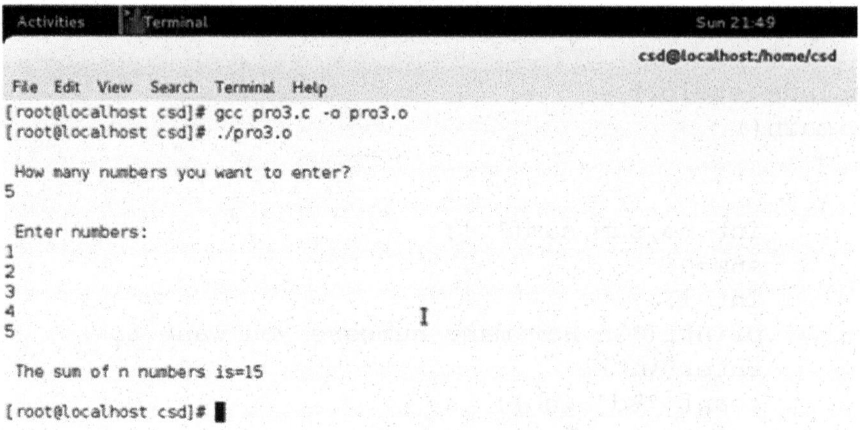

Figure 2.30   Output screen of Program 2.30.

```c
/* Multiplication table of a given number */
#include<stdio.h>
int main()
{
        int i,n,b;
        printf("\n Enter any number\n");
        scanf("%d",&n);
        printf("\n The multiplication table for %d
        is\n\n",n);
        for(i=1;i<=10;i++)
        {
                b=n*i;
                printf("%d\n",b);
        }
        return 0;

}
```

Here, user is prompted to enter the number for which multiplication table has to be displayed. User has entered 2 as shown in Figure 2.31, and multiplication table of 2 is displayed.

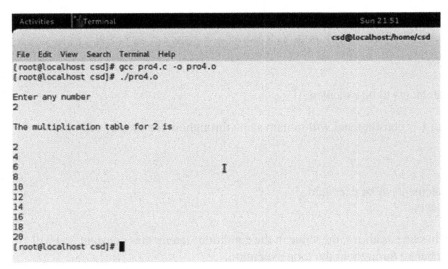

Figure 2.31   Output screen of Program 2.31.

2.9 Loop Control Statements

Sometimes we may want to leave the loop before its completion; then, we use these loop control statement. Program's normal execution sequence can be altered (changed) by using these statements. If some variables are declared and created in this scope, then they are destroyed automatically as soon as execution leaves the scope.

Following control statements are supported by C.

Break statement: This statement breaks the loop or switch statement. It is useful if we want to break out of any loop in between without executing it completely. Once break statement is executed, loop is terminated and control is returned to statement immediately following the loop.

Continue Statement: This statement helps in skipping some statements within the loop body. Some subcondition may not require the whole body of the loop to be executed; then, in that case, continue statement can be inserted above the statements which need to be skipped from execution.

2.10 Infinite Loop

If the condition statement of the loop always evaluates to be true, then that loop never ends and becomes an infinite loop. Here, break statement can be used to break away from such loops. Here while, do while, and for loops can be used to make endless loops as follows:

while(1)

{Statements to be executed.. }

Here, 1 is constant and will remain same throughout the loop execution.

do

{ statements to be executed..}
while(1);

Again same as above, the value in the condition statement is constant and will not change throughout the loop execution.

for(int i=0; ; ++i)

{ statements to be executed..}

The for loop initializes the variable and increments its value but there to condition check to terminate the loop, so it will execute infinitely.

Program 2.32: Example of a program that never ends

Here, for loop is initialized and incremented, but condition for loop termination is not given. So program executes infinitely.

```
/* Program with infinite loop */

#include<stdio.h>
int main()

{
        int i;
        printf("\n Enter any number...\n");
        for(i=1;;i++)
        {
                scanf("%d",&i);
```

```
            printf("\n Entered number is:%d\n",
            i);
    }
    printf("\n\n");
    return 0;
}
```

Here, as shown in Figure 2.32 as soon as the program executes, user is prompted to enter the number and the infinite loop starts, which is then terminated abruptly.

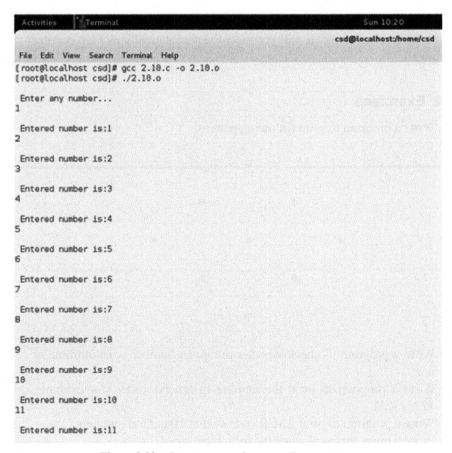

Figure 2.32   Output screen of never-ending program.

2.11 Conclusion

In this chapter, how to write C programs using control structures and looping constructs was studied. Now, one can make decisions and take appropriate action regarding the problem. As per given syntax, it is possible to write it in the form of C programs. The syntax and example programs are provided in to Linux environment. So, one can do C programming in Linux environment easily. The branching statements give alternate solutions to the problem. Also, it increases the ability to think over problem in multiple directions. As the number of ways of solution increases, branches in program statement increase. Each step in branching statement gives new alternative solution and makes program more structured and easy to read. Looping constructs perform same task number of times. It saves writing same code statement again and again for specific number of times. It will get executed automatically as loop counter increments.

2.12 Exercises

1. Write a program to print following pattern.

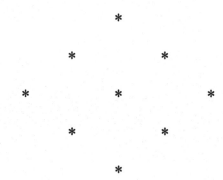

2. Write a program to check whether the given number is palindrome or not.
3. Write a program to print the number in reverse order (for example, 123 = 321).
4. Write a program to print ASCII code of first 10 natural numbers.
5. Write a program to calculate the following series.
   1.0 + 1.1 + 1.2 + 1.3 +........+ up to n terms.

3

Arrays and Pointers

3.1 Why Array?

Variables are used to store the values in the memory. If we want to store student roll number, then we need to declare one variable of type integer to store it. Similarly, if we need to store roll numbers of all the students in the class having 30 strength, then we may need to declare type integer of size 30. This is the right way but not the proper way to write a program. Here, it will increase the number of code lines as well as the complexity of program reading. Each time we need to remember all the variable names. We need to keep track of these variables in the program. The memory allocation for these variables may or may not be contiguous. So, we should declare integer array of size 30 to store roll numbers. By doing so, we can reduce the number of lines of code. One array name along with the index number is sufficient to handle all the roll numbers.

3.1.1 Features

Array has the following features:

- Array can store only one type of variable, that is, variable having similar data type.
- Array supports contiguous memory allocation.
- Array is a static and linear data structure.
- Allocated memory is fixed in size once an array is declared.
- Array elements can be accessed using index number.
- First element of the array is always at 0th position. If we store 10 elements in an array, then the index number starts from 0th position to 9th.
- Arrays can be multidimensional.

3.1.2 Definition

Array is a finite collection of data items having similar data type. Array elements are stored in the memory in a contiguous fashion. If we want to store data of similar type, then we should use array. If we want to display student roll number in a class, then we can display it using array.

3.2 Types of Arrays

Array has the following types. They are different in their dimensions in which data are stored in the memory.

3.2.1 Single-Dimensional Array

Single-dimensional array stores elements in one dimension only.

Declaration

data_type Array_name[n] ;
where data_type is any valid C data type
    Array_name is the name of an array
    n in [] describes the number of elements to be stored in an array

Example

int num[10] is a single-dimensional array.

- int num[10] ;
  array num will store 10 integer numbers.
- int num[10]={1,2,3,4,5,6,7,8,9,10}
  array num stores 10 integer numbers starting from 1 to 10.

Memory Representation

The memory representation for above integer array num[10] is shown in Table 3.1.

Table 3.1 Memory representation of single dimension array

Value	1	2	3	4	5	6	7	8	9	10
Address	1001	1003	1005	1007	1009	1011	1013	1015	1017	1019

Program 3.1: Program to calculate the sum of odd and even numbers in a set using one-dimensional array

```c
/* Program to calculate the sum of even numbers and
odd numbers in a set of values */

#include<stdio.h>
int main()

{
        int A1[5],v,sum1,sum2;
        sum1=0;
        sum2=0;
        printf("\n Enter the elements:\n");
        for(v=0;v<5; ++v)
        {
                scanf("%d",&A1[v]);
        }
        for(v=0;v<5;++v)
        {
                if(A1[v]%2==0)
                        sum1=sum1+A1[v];
                else
                        sum2=sum2+A1[v];
        }
        printf("\n Sum of even numbers is:=%d\n\n",
        sum1);
        printf("\n Sum of odd numbers is:=%d\n\n",
        sum2);
        return 0;

}
```

Here, in the above program, we have declared one-dimensional integer array having 5 elements. *i* is an integer type variable used to traverse the array elements. Using for loop, we can traverse the array. We required one for loop to input the array elements and one more for loop to output the array elements. If we have to process the array elements, then we need to declare one for loop to perform the same task.

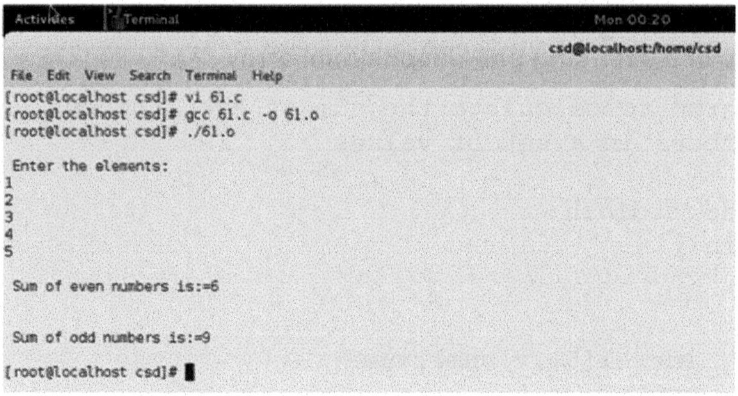

Figure 3.1 Output screen of Program 3.1.

Here, as shown in Figure 3.1 the user is asked to enter the elements for set. The program then checks for odd and even numbers and sum them separately. Similar program has already been discussed in Program 2.26 in Chapter 2.

Program 3.2: Program to store a set of numbers using one-dimensional array

Here, the user is asked to enter ten integer numbers. Then, the program checks the number greater than zero, less than zero, and equal to zero; accordingly, the count for each type is incremented. Similar program has already been discussed in Program 2.27 in Chapter 2 with focus on for loop.

```c
/* Program to count positive integers, negative
integers, and zeros in a given set of integers */

#include<stdio.h>
int main()
{
        int A1[10],i,cnt1,cnt2,cnt3;
        cnt1=0;
        cnt2=0;
        cnt3=0;
        printf("\n Enter Array elements\n");
        for(i=0;i<10;++i)
        {
                scanf("\n %d",&A1[i]);
```

```
}
for(i=0;i<10;++i)
{
        if(A1[i]>0)
                cnt1=cnt1+1;
        else if(A1[i]<0)
                cnt2=cnt2+1;
        else if(A1[i]==0)
                cnt3=cnt3+1;
}
printf("\n The positive numbers are:=%d\n\
n",cnt1);
printf("\n The negative numbers are:=%d\n\
n",cnt2);
printf("\n The number of zeros:=%d\n\n",
cnt3);
return 0;
}
```

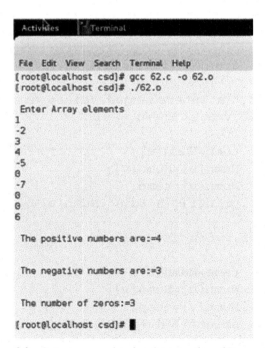

Figure 3.2 Output screen showing the execution of Program 3.2.

Here, the user is asked to enter integer numbers. As shown in Figure 3.2, 10 numbers are entered. After checking for positive, negative, and zero integer counts, the output is displayed.

Program 3.3: Program used to reverse the contents of one-dimensional array

Here, the user enters 10 values. To reverse the content of this array, the first value is swapped with the last value in the array and the second value is swapped with the second last value in the array and so on. The for loop at the last in the program prints the details of swapping.

```c
/* Program to interchange values in an array */

#include<stdio.h>
int main()
{
        int Numn[10],u,v,temp;
        printf("\n Enter Elements\n");
        for(u=0;u<10;++u)
        {
                scanf("%d",&Numn[i]);
        }
        printf("\n Interchanged Values are:\n\n");
        for(u=0,v=9;u<5;++u,--v)
        {
                temp=Numn[u];
                Numn[u]=Numn[v];
                Numn[v]=temp;
                printf("\n %d\n",Numn[u]);
        }
        for(u=5,v=0;u<10;++u,++v)
        {
                temp=Numn[u];
                Numn[u]=Numn[v];
                Numn[v]=temp;
                printf("\n %d \n",Numn[v]);
        }
```

```
printf("Format second\n");
for(u=0,v=9;i<5;++u,--v)
{
        printf("\n %d- %d ",Numn[u],Numn[v]);
}
return 0;
}
```

Here, as shown in Figure 3.3 it can be observed that user enters 10 numbers from 1 to 10 in order. Results after interchanging are displayed showing numbers from 10 to 1 in order, thus reversing the sequence. It also shows which numbers are swapped.

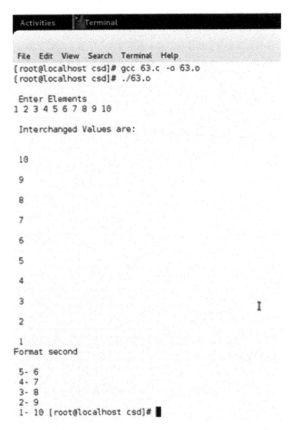

Figure 3.3 Output screen of Program 3.3 after execution.

Program 3.4

Here, two arrays are entered by the user, each of size 5. These arrays are then
sorted in ascending order using simple sorting method (bubble sort). After
sorting, they are merged into a single large array in ascending order.

```c
/* Program to merge two sorted arrays */

#include<stdio.h>
int main()

{
        int A1[5],i,j,temp,B1[5],C1[10],k;
        printf("\n Enter the elements of Array1\n");
        for(i=0;i<5;++i)
        {
                scanf("%d",&A1[i]);
        }
        for(i=0;i<5;++i)
        {
                for(j=i+1;j<5;++j)
                {
                        if(A1[i]>A1[j])
                        {
                                temp=A1[i];
                                A1[i]=A1[j];
                                A1[j]=temp;
                        }
                }
        }


        printf("\n Sorted elements in Array1 :\n");
        for(i=0;i<5;++i)
        {
                printf("%d\t",A1[i]);
        }
        printf("\n Enter the values of Array2");
        for(i=0;i<5;++i)
```

```
{
            scanf("%d",&B1[i]);
}
printf("\n Sorted elements in array2 :");
for(i=0;i<5;++i)
{
            for(j=i+1;j<5;++j)
            {
                        if(B1[i]>B1[j])
                        {
                                    temp=B1[i];
                                    B1[i]=B1[j];
                                    B1[j]=temp;

                        }

            }

}
for(i=0;i<5;++i)
{
            printf("%d\t",B1[i]);

}
printf("\nThe Merged elements of Array1 and
Array2:\n");
for(k=0;k<10;++k)

{
            for(i=0,k=0;i<5,k<5;++i,++k)

            {
                        C1[k]=A1[i];
            }
            for(j=0,k=5;j<10,k<10;++j, ++k)
            {
                        C1[k]=B1[j];
            }
}
for(k=0;k<10;++k)
{
            printf("%d\t",C1[k]);
}
```

```
printf("\n Sorted Array3 is:\n");
for(i=0;i<10;++i)
{
          for(j=0;j<10;++j)
          {
                    if(C1[i]<C1[j])
                    {
                              temp=C1[i];
                              C1[i]=C1[j];
                              C1[j]=temp;
                    }
          }
}
for(i=0;i<10;++i)
{
          printf("%d\t",C1[i]);
}
printf("\n\n");
}
```

As shown in Figure 3.4, user is asked to enter the elements for first array. These elements are then sorted and displayed. Then, the user is asked to enter the elements for second array which are also arranged in ascending order. These two sorted arrays are merged to form a large array which is also in ascending order.

3.2.2 Two-Dimensional Array

It is also possible to store the array elements in two dimensions. These are array of arrays. It is also called as matrix.

Declaration

data_type Array_name[n][n] ;
where data_type is any valid C data type
      Array_name is the name of an array
      n in [] describes the number of elements to be stored in an array
      Here, first [n] describes the row elements and second [n] describes the column elements.

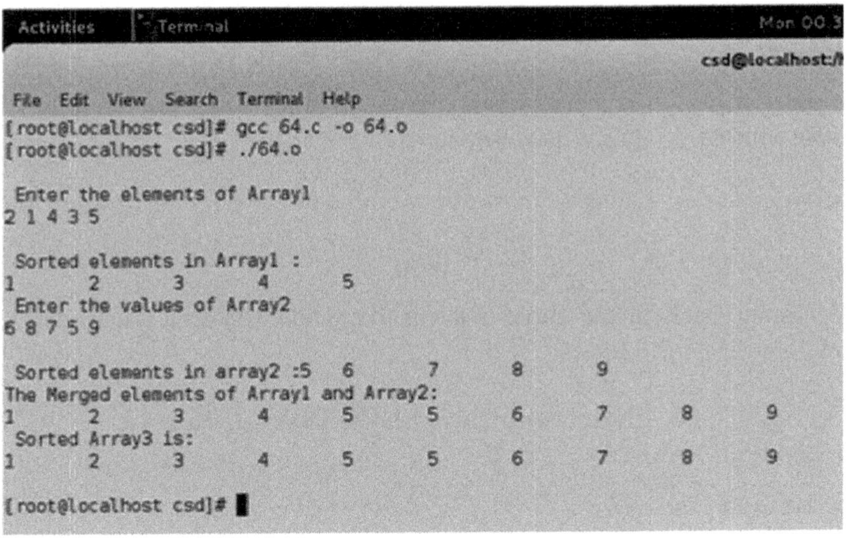

Figure 3.4 Output screen of Program 3.4 after execution.

Example

int num[2][2] is a two-dimensional array.

Here, first subscript can be row elements and second subscript can be column elements.

int Num[2][2]={{7,8},{9,10}}

Memory Representation

Table 3.2 Memory representation of double dimension array

Num[2][2]	Column 0	Column 1
Row 0	7	8
Row 1	9	10

Element	Num[0][0]	Num[0][1]	Num[1][0]	Num[1][1]
Value	7	8	9	10
Address	5001	5003	5005	5007

Program 3.5: Program to check whether the matrix is diagonal or not

A diagonal matrix is the matrix whose all non diagonal entries are zero. For example a matrix A[3][3] is diagonal matrix.

$$A[3][3] = \begin{matrix} 4 & 0 & 0 \\ 0 & 2 & 0 \\ 0 & 0 & 7 \end{matrix}$$

This program checks all the entries of the matrix to find whether it is diagonal or not.

```c
/* Program to check whether the entered matrix is
a diagonal matrix or not */

#include<stdio.h>
int main()
{
        int A1[5][5],i,flag=0,j;
        printf("\n Enter the Matrix :");
        for(i=0;i<5;++i)
        {
                for(j=0;j<5;++j)
                {
                        scanf("%d",&A1[i][j]);
                }
        }
        for(i=0;i<5;++i)
        {
                for(j=0;j<5;++j)
                {
                        if(i!=j)
                        {


                                if(A1[i][j]==0)
                                flag=flag+1;
                        }
                }
        }
```

```
            if(flag==5*(5-1))
                    printf("\n The given matrix is
                    diagonal matrix\n\n");
            else
                    printf("\n The given matrix is not
                    diagonal matrix\n\n");
            return 0;
}
```

As shown above, here the user is asked to enter the elements for a 5 × 5 square matrix. After processing the elements, it displays the matrix as all the elements in the diagonal are nonzero and rest of the elements are zero. Figure 3.5 displays the details of output.

Program 3.6: Program to add row elements, column elements, and diagonal elements of a matrix

This program consists of four for loops. The first for loop is to accept the matrix elements. The second for loop is to sum the row elements, the third is to sum the column elements, and the last is to sum the diagonal elements of the matrix.

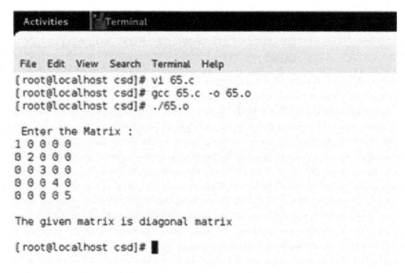

Figure 3.5 Output screen of Program 3.5.

```c
/* Program to calculate the row sum, column sum, and
diagonal sum of a given matrix */

#include<stdio.h>
int main()
{
        int A1[5][5],r,c,m,n,sum=0;
        printf("\n Enter Matrix :\n");
        for(m=0;m<5;++m)
        {

                for(n=0;n<5;++n)
                {
                        scanf("%d",&A1[m][n]);
                }


        }
        for(m=0;m<5;++m)
        {
                for(n=0;n<5;++n)
                {
                        r=A1[m][n];
                        sum=sum+r;
                }
                printf("\n Sum of Row %d =%d\n",u,
                sum);
                sum=0;

        }
        for(n=0;n<5;++n)
        {

                for(m=0;m<5;++m)
                {
                        c=A1[m][n];
                        sum=sum+c;
                }
                printf("\n Sum of Column %d=%d\n",
                v,sum);
                sum=0;
```

```
}
for(m=0;m<5;++m)
{
        for(n=0;n<5;++n)
        {
                if(m==n)
                        sum=sum+A1[m][n];
        }

}
printf("\n Sum of Diagonal is:%d\n\n",sum);
return 0;
}
```

As shown in Figure 3.6, the user is asked to enter the data for the matrix of size 5 X 5. As the elements entered are same in all the rows, it gives the same total, i.e., 15 for each row.

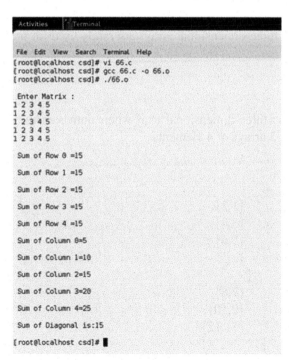

Figure 3.6 Output screen of Program 3.6.

Column entries are different for different columns, so we get a different total for each column. As there are same entries for each row, diagonal total is also 15.

3.2.3 Three-Dimensional Array

Three-dimensional arrays are rarely used in practice. They are array of arrays of arrays.

Declaration

data_type Array_name[n][n][n] ;
where data_type is any valid C data type
    Array_name is the name of an array
    n in [][][] describes the number of elements to be stored in an array

The array is a static data structure. The memory allocated using array is fixed, and we cannot change the size of the array once they are declared. So, if we declare the array of size 10, then only 5 elements are stored in the memory, and the rest of the memory space is wasted and no longer be used. So, in real-life situation, three-dimensional arrays are rarely used for programming purpose.

Example

int num[2][3][4] is a three-dimensional array where num is an array consisting of 2 arrays having 3 arrays of 4 elements.

```
int num[2][3][2]= {
                  {
                        {1,2},
                        {3,4},
                        {5,6}
                        },
                  {
                        {7,8},
                        {9,10},
                        {11,12}
                  }
            }
```

3.3 Pointer Data Type

Pointer variables are special variables used to store the address of another variable of any data type. We can point to memory location by using pointer variable. The memory locations are always numbers. They are integers. So, pointer variable always contains integer values, known as whole numbers.

The following program demonstrates how to print the address of a variable without using pointer variable.

Program 3.7: Program to print the address of a variable without using pointer

This program prints the address of a variable in the decimal form. Here, %u is for unsigned decimal number. It prints the offset address but not the code segment address.

```
/* Program to demonstrate how to print the address
of a variable without using the pointer variable */

#include<stdio.h>
int main()
{

        int x=684;
        printf("\n The value of integer variable x
        is :%d\n",x);
        printf("\n The Address of an integer
        variable x in memory is: %u\n\n",&x);
        return 0;

}
```

Here, as shown in Figure 3.7 the integer value 684 is stored in the variable in the program itself, so no user interaction. In the above program, the format specifier %u prints the unsigned integer address of a variable. The & operator is a "address of" operator, and we use it with scanf() function all the time.

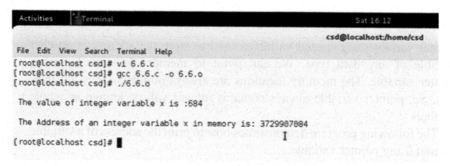

Figure 3.7   Output screen of Program 3.7.

3.3.1 Pointer Declaration

Pointers are declared in the same way as any other variable but the only difference is presence of '*' in declaration.

int x=10;

The above declaration says that memory space of 2 bytes having the name 'x' is reserved and the value stored at that location is 10.

int *ptr;

Here, ptr is a pointer variable which will store the address of an integer variable.

char *ptr;

Here, ptr is a pointer variable which will store the address of a character variable.

float *ptr;

Here, ptr is a pointer variable which will store the address of a float variable.

int **ptr1;

Here, ptr1 is a pointer variable which will store the address of a pointer variable.

3.3.2 Pointer Initialization

Once the pointer is declared it should be pointing to some memory location which is done using pointer initialization.

ptr = &x;

Here, variable ptr will hold the address of x.

The meaning of '&' operator is "address of" and that of '*' operator is "value at address."

Program 3.8: Program to show how to use pointers pointing to different data types

An integer pointer can point to an integer variable, a character pointer can point to a character variable, and a float pointer can point to a float variable. Here, pointer data type depends on what data type it is pointing to. A void pointer can pointto any data type.

```c
/* Program to demonstrate the use of a pointer
variable */

#include<stdio.h>
int main()
{
        int i,*p,**ptr;
        float j,*q;
        char k,*r;
        i=684;
        j=25.6;
        k='s';
        p=&i;
        q=&j;
        r=&k;
        ptr=&p;
        printf("\n The value of an integer variable
        'i' is:%d\n",i);
        printf("\n The address of an integer
        variable 'i' in memory is:%u\n",p);
        printf("\n The value of a float variable 'j'
        is:%.2f\n",j);
        printf("\n The address of an integer
        variable 'j' in memory is:%u\n",q);
        printf("\n The value of a character variable
        'k' is:%c\n",k);
        printf("\n The address of a character
        variable 'k' in memory is:%u\n",r);
```

```
        printf("\n The value at %u memory location
        is :%d\n",p,*p);
        printf("\n The value at %u memory location
        is :%.2f\n",q,*q);
        printf("\n The value at %u memory location
        is :%c\n",r,*r);
        printf("\n The value of pointer variable
        'ptr' is :%u\n",ptr);
        printf("\n The address of a pointer variable
        'p' in memory is:%u\n",ptr);
        printf("\n The value at %u memory location
        is :%u\n\n",ptr,*ptr);
        return 0;
}
```

In the above program, we have three types of pointers pointing to their respective data types, viz. integer, character, and float. There is one more variable **ptr, and this variable holds the address of pointer ptr. Figure 3.8 shows the values of the variables and their addresses in the memory.

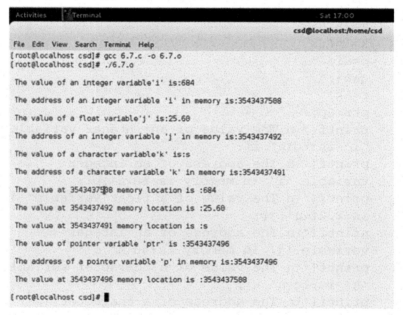

Figure 3.8 Output screen of Program 3.8.

3.4 Arrays and Pointers

Pointer plays a very important role in arrays. Before going into detail how pointers are used in arrays, we will first see the pointer arithmetic.

int *ptr;   // integer pointer
ptr++;   // pointer points to the immediate next memory location from the current memory location
ptr--;   // pointer points to the immediate back memory location from the current memory location
ptr=ptr+2;   //pointer points to the next second memory location from the current memory location
ptr=ptr-2;   //pointer points to the back second memory location from the current memory location

Program 3.9: Program to show the valid arithmetic operations on pointers

Like any normal variable, pointer variables can be incremented or decremented using ++ and - - operators. Pointer values can be added or subtracted, but two addresses cannot be added or subtracted using + and - operators. The following program demonstrates valid operations on pointers.

```
/* Program to demonstrate the pointer arithmetic */

#include<stdio.h>
int main()
{
        int i,*ptr,j,*ptr1,*pt,*pt1;
        ptr=&i;
        ptr1=&j;
        i=10;
        j=20;
        int a[3]={1,2,3};
        printf("\n Values of i and j are :%d and
        %d\n",i,j);
        printf("-----------------------------------
        ----------------");
        printf("\n The address of 'i' is:%u\n",
        ptr);
```

```
printf("\n The address of 'j' is:%u\n",
ptr1);
printf("--------------------------------
----------------");
printf("\n The result of '*ptr1-*ptr' is:
%d",*ptr1-*ptr);
printf("\n--------------------------------
------------------");
ptr++;
printf("\n The value of pointer variable
'ptr' after incremented by one is:%u\n",
ptr);
ptr=ptr+2;
printf("\n The value of pointer variable
'ptr' after incremented by two is:%u\n",
ptr);
ptr=ptr--;
printf("\n The value of pointer variable
'ptr' after decremented by one is:%u\n",
ptr);
ptr=ptr-3;
printf("\n The value of pointer variable
'ptr' after decremented by three is:%u\n",
ptr);
printf("\n--------------------------------
------------------");
printf("\n The array elements in array 'a'
are :\n");
for(i=0;i<3;++i)
{

        printf("\t %d",a[i]);

}

printf("\n The addresses of array elements
in array 'a' are :\n");
for(i=0;i<3;++i)

    {
```

```
                    printf("\t %u",pt);
                    pt++;
    }
    printf("\n------------------------------
    --------------------");
    pt=&a[0];
    pt1=&a[2];
    printf("\n The result of 'pt1-pt' is :%d\n
    \n",pt1-pt);
    printf("\n------------------------------
    --------------------");
    pt=&a[1];
    pt1=(a+1);
    printf("\n If the expressions are : pt=&a
    [1] and pt1=(a+1) then...\n");
    if(pt==pt1)
                    printf("\n The two pointers 'pt'
                    and 'pt1' are pointing to the
                    same location\n\n");
    else
                    printf("\n The two pointers 'pt'
                    and 'pt1' are not pointing to
                    the same location\n\n");
    return 0;
}
```

As shown in Figure 3.9, two variables and their addresses are stored in the memory.

After performing subtraction on pointers pointing at them, it displays the result as 10. Also, increment and decrement operations are shown. The last part of the output shows relationship between arrays and pointers.

3.4.1 Pointers and One-Dimensional Arrays

Pointer variable can be used to traverse the one-dimensional array elements easily. Also, we can pass the entire array to a function as a parameter using pointer.

We can traverse the array elements using C pointer arithmetic.

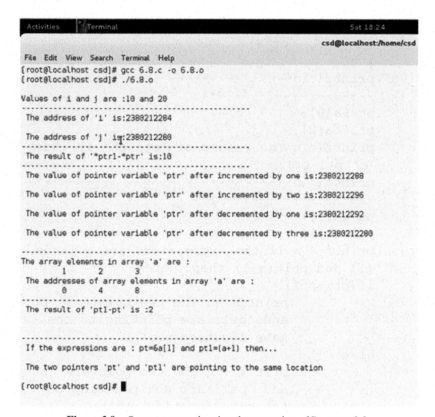

Figure 3.9 Output screen showing the execution of Program 3.9.

Program 3.10: Program to use the pointer to print the contents of a one-dimensional array

Here, a pointer is made to point to the base address of the array by storing the address of the first element of array in it. Pointer arithmetic plays an important role here. As shown in the program below, j is the pointer variable which stores the base address of the array and is incremented to point to the next element in the array.

```
/* Program to demonstrate 1-D array with function
using pointer */

#include<stdio.h>
int display(int *j,int n);
```

```
int main()
{

        int Y[5],i;
        printf("\n Enter Values:\n");
        for(i=0;i<5;++i)
        {

                scanf("%d",&Y[i]);
        }

        display(&Y[0],5);
}

int display(int *j,int n)

{
        int x;
        printf("\n The elements in 1-D Array are:
        \n\n");
        for(x=0;x<n;++x)
        {
                printf("%d\n",*j);
                j++;
        }
        return 0;

}
```

As shown in Figure 3.10, the user is asked to enter the array elements. These elements are then printed using pointer variable. Array is traversed using pointer instead of index.

3.4.2 Pointers and Two-Dimensional Arrays

It is also possible to access the elements of a two-dimensional array using pointers. Two-dimensional arrays are array of arrays. We can point to the address of each one-dimensional array in a two-dimensional array.

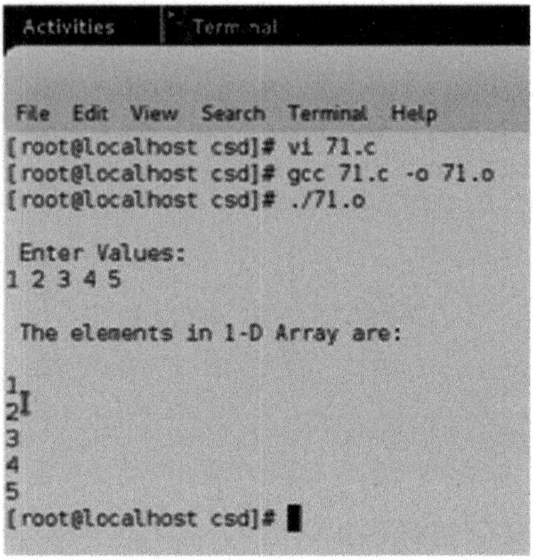

Figure 3.10   Output screen of Program 3.10.

Program 3.11: Program to show the use of pointers with two-dimensional array

Here, the contents of array are printed using three different functions. Here, disp1 and disp2 has pointer as the parameter for printing and disp3 is a display function without any pointer parameter.

```
/* Program to demonstrate 2-D array with function
using pointer */

#include<stdio.h>
int Arr1[4][2]={{1,2},{2,3},{3,4},{4,5}};
int disp1(int *p,int,int);
int disp2(int (*p)[2],int,int);
int disp3(int s[][2],int,int);

int disp1(int *p,int Row,int Column)
{
        int i,j,*pt;
        for(i=0;i<Row;++i)
```

```
        {
                        for(j=0;j<Column;++j)
                        {
                                printf("%d",*(p+i*Column
                                +j));
                                pt++;
                        }
                        printf("\n");
        }

}

int disp3(int Arr1[][2],int Row1, int Column1)

{
        int i,j;
        for(i=0;i<Row1;++i)
        {
                        for(j=0;j<Column1;++j)
                        {
                                printf("%d",Arr1[i][j]);
                        }
                        printf("\n");
        }
}

int disp2(int (*p)[2],int Row1,int Column1)

{
        int i,j,*pt;
        pt=p;
        for(i=0;i<Row1;++i)
        {
                        for(j=0;j<Column1;++j)
                        {

                                printf("%d",*pt);
                                pt++;
                        }
```

```
                    printf("\n");
        }
}
int main()
{
        int R,C;
        R=4;
        C=2;
        disp1(Arr1,R,C);
        printf("\n");
        disp2(Arr1,R,C);
        printf("\n");
        disp3(Arr1,R,C);
        printf("\n");
        return 0;
}
```

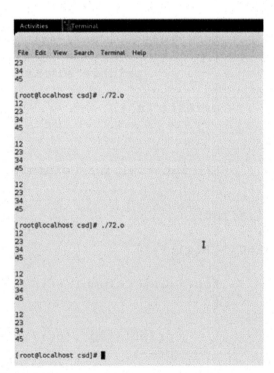

Figure 3.11   Output screen of Program 3.11 after execution.

It is observed in Figure 3.11 that there is no user interaction. Array is initialized in the code itself. The contents of the initialized array are displayed using different display functions.

Program 3.12: Program to show the use of array of pointers

Here, A[] is the array of pointers, and b[] is a simple integer array. In the first for loop, elements are added to array b[], and the addresses of these elements are added to array A, which is the array of pointers. The second for loop displays the address of array elements in the decimal form using %u.

```
/* Program to demonstrate array of pointers
containing the address of another array elements */

#include<stdio.h>
int main()

{

        int *A[5],b[5],i;
        printf("\n Enter Array Elements:");
        for(u=0;u<5;++u)
        {
                scanf("%d",&b[u]);
                A[u]=&b[u];
        }
        printf("\n The Address of an Array elements
        is:\n");
        for(u=0;u<5;++u)
        {
                printf("%u\n",A[u]);
        }
        return 0;

}
```

Here, as shown in Figure 3.12 the user is asked to enter the elements for the array. Addresses of these elements are stored in another array which is the array of pointers, and the same is displayed.

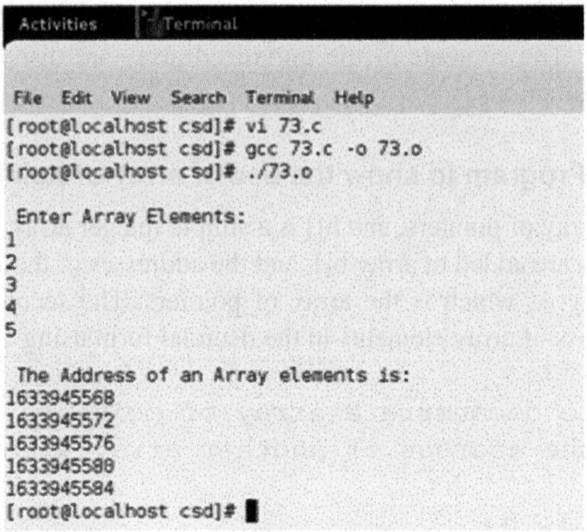

Figure 3.12 Output screen of Program 3.12 after execution.

Program 3.13: Program to access the array elements using pointer

As shown in Program 3.12, the address of array elements is stored in the array of pointers, i.e., *A[]. This array (*A[]) is used to access the elements stored at that address. The values of these elements are added to compute the arithmetic mean.

Figure 3.13 Output screen of Program 3.13.

```
/* Program to calculate the arithmetic mean of array
elements using pointer */

#include<stdio.h>
int main()
{
        int *A[5],b[5],sum,mean,i;
        sum=0;
        printf("\n Enter Values:\n");
        for(u=0;u<5;++u)
        {
                scanf("%d",&b[u]);
                A[u]=&b[u];
        }
        for(u=0;u<5;++u)
        {
                sum=sum+*A[u];
                mean=sum/5;
        }
        printf("\n Mean=%d\n\n",mean);
        return 0;
}
```

The user is asked to enter values for array elements as shown in Figure 3.13. These values are summed and divided by the number of elements in the array to get the mean. Values are accessed using pointer.

3.5 Conclusion

In this chapter, we have studied how to organize data using array. Now, one can work with multiple data items of same data type using array. Also, we have seen how to store data in one dimension as well as multiple dimensions. Two-dimensional arrays are known as matrices. Data are organized in the form of rows and columns. Three-dimensional arrays are rarely used. There are some limitations in using array. As we have seen arrays are static, there is waste of memory if some data items are not allocated at their positions. Memory is reserved such that the unallocated space cannot be used for other purposes. Also, we have studied pointer data type here. Along with pointer basics, pointer arithmetic and its use with array while doing programming have been studied. Pointers are so useful in arrays while accessing data items. Though programming with pointer is a difficult task, pointers are most useful in programming constructs and it is a key feature of C programming.

3.6 Exercises

1. Write a program to insert book details in a library.
2. Write a program to input salary of employees in a company and display the average salary.
3. Write a program to display parking details of 10 cars and search car details according to the entered car number.
4. Write a program to perform arithmetic operations on a matrix.
5. Write a program to store names of 10 students in an array and sort them in an ascending order.

4

Programming for Functional Functions

4.1 Introduction

Functions are block of code enclosed within pair of curly braces. One can use functions to call code block multiple times in the program.

Functions return something to the program or user. They have return value including void data type.

General form:

```
return_type function_name(data_type variable1, data_type variable2,...)
{
    //statements
}
```

The above syntax defines the prototype for the function.

Here,

return_type is any valid C data type,

function_name is user defined function name, and

variable1, variable2,... are the variables used as function parameters. The function parameters are optional. The function can be having zero input parameters by keeping ( ) parenthesis empty.

4.2 Function Declaration

Function declaration defines function prototype as follows.
return_type function_name(data_type variable1, data_type variable2,....);

Example:

int addition(int a, int b);

4.3 Function Definition

Function definition defines function body where one can write the function code. Function body always enclosed within pair of opening and closing curly braces.

return_type function_name(data_type variable1, data_type variable2,....)
{
    // statements
}

Example:

int addition(int a, int b)
{
    printf("\n addition is :%d",a+b);
}

4.4 Function Call

Function call is used to pass control to the function along with arguments if any. It does not include return type of the function but only function name with arguments. Following is the syntax for function call.

function_name(variable1,variable2,...);

Example:

addition(5,2);

Program 4.1: To greet user

Here, a function Tmsg() is used. This function does not return any value, and return type is void and does not have any parameters passed to it. This type of function does not manipulate data but works as a read-only function. This function is called within main function for execution.

```
/* Program to display text message using function */

#include<stdio.h>
void Tmsg();
int main()
{
        Tmsg();
        return 0;
}

void Tmsg()
{
        printf("\n Hello!Have a Nice Day!!!\n\n");

}
```

Here, as soon as the program executes, it greets user with a message as shown in Figure 4.1. There is no user interaction. Message is built inside the code itself.

Here, in the above program, we have declared message() function to display a message. The function just displays a message and it does not return anything so; return type of function is void and also it does not accept any parameters, so the function is having empty parenthesis. The main() returns true value always, so the return type of main() is int.

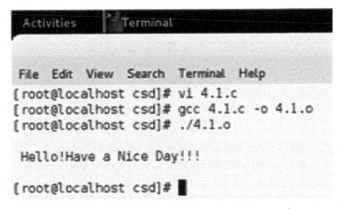

Figure 4.1 Output screen of Program 4.1 after execution.

Program 4.2: Computing factorial of a positive integer

Details about factorial have been discussed in Chapter 2. Here, factorial of a given integer is computed using a function. An integer whose factorial has to be computed is passed to a function as an argument, and function computes the factorial and returns the factorial value.

```c
/* Program to calculate factorial of a given number
using function */

#include<stdio.h>
int factorial(int a);
int main()
{
        int no,f;
        printf("\n Enter any number\n");
        scanf("%d",&no);
        f=factorial(no);
        printf("\n The factorial of given number is=
        %d\n\n",f);
        return 0;
}
int factorial(int a)
{
        int fact;
        fact=1;
        while(a>0)
        {
                fact=fact*a;
                a--;
        }
        return fact;
}
```

User is prompted to enter any integer number as shown in Figure 4.2. Factorial of the number is computed and displayed. Data type of the variables in the program is integer, which can support the factorial value only for limited numbers as the value increases exponentially for larger number, long integer, double, etc. will be required.

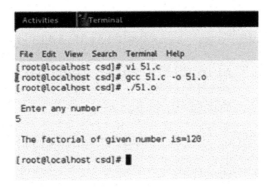

Figure 4.2 Output screen of Program 4.2 after execution.

Program 4.3: To find the highest three test and assignment marks of students

Here, data of three students is collected in an array. There are two arrays used: one to store assignment marks, a[5]; and second to store test marks, t[5]. Using nested for loops, data is entered into respective arrays. The best three assignment marks along with their averages are displayed for each student. Similarly, the best three test marks and averages of test marks are also displayed.

```
/* Program to find highest score of a student using
function */

#include<stdio.h>
int best(int *a,int *b,int n);
int big(float x);
int tavg();
int cnt=0;
int i,j,k,temp;
float avg[3];
int main()
{
        int a[5],t[5],i,j,*p,*q;
        for(i=0;i<3;i++)
        {
                printf("\n Enter assignment marks
                for student %d:\n",i+1);
                for(j=0;j<5;j++)
```

```
                    {
                            scanf("%d",&a[j]);
                    }
                    printf("\n Enter test marks for
                    student %d:\n",i+1);
                    for(j=0;j<5;j++)
                    {
                            scanf("%d",&t[j]);
                    }
                    p&a[0];
                    q=&t[0];
                    best(p,q,5);
            }
        tavg();
}
int best(int *a,int *b,int n)
{
        int c[5],d[5],sum=0,sum1=0;
        float avg1,avg2,tavrg;
        for(i=0;i<n;i++)
        {
                c[i]=*a;
                a++;
                d[i]=*b;
                b++;
        }
        for(j=0;j<5;j++)
        {
                for(k=j+1;k<5;k++)
                {
                        if(c[j]<c[k])
                        {
                                temp=c[j];
                                c[j]=c[k];
                                c[k]=temp;
                        }
                        if(d[j]<d[k])
                        {
                                temp=d[j];
```

```
                                d[j]=d[k];
                                d[k]=temp;
                        }
                }
        }
        printf("\n Best Three assignment marks are:
        \n");
        for(i=0;i<3;i++)
        {
                printf("\n %d\n",c[i]);
                sum=sum+c[i];
        }
        printf("\n Sum of best three assignment
        marks is =%d\n",sum);
        avg1=sum/3;
        printf("\n Average of best assignment marks
        is:%.2f\n",avg1);
        printf("\n Best three test marks\n");
        for(i=0;i<3;i++)
        {
                printf("%d\n",d[i]);
                sum1=sum1+d[i];
        }
        printf("\n Sum of best three test marks is
        =%d\n",sum1);
        avg2=sum1/3;
        printf("\n Average of Best test marks is =
        %.2f\n",avg2);
        tavrg=(avg1+avg2)/2;
        printf("\n Total average=%.2f\n",tavrg);
        big(tavrg);
        return 0;
}
int big(float x)
{
        avg[cnt]=x;
        cnt++;
        return 0;
}
int tavg()
```

```
{
        for(i=0;i<3;i++)
        {
                for(j=i;j<3;j++)
                {
                        if(avg[i]<avg[j])
                        {
                                temp=avg[i];
                                avg[i]=avg[j];
                                avg[j]=temp;
                        }
                }
        }
        for(i=0;i<1;i++)
        {
                printf("\n The Biggest average is:
                %f",avg[i]);
        }
        return 0;
}
```

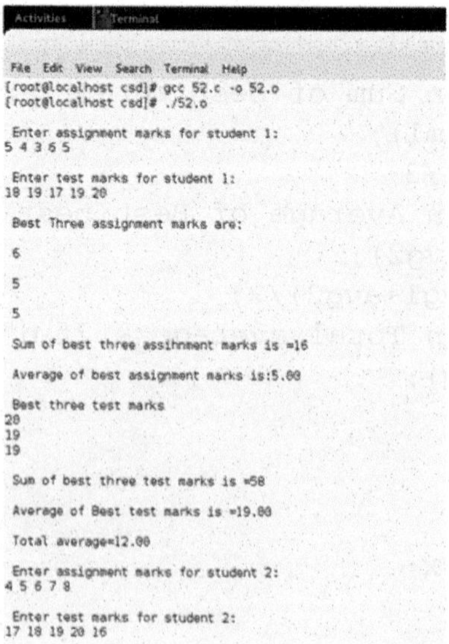

Figure 4.3   Output screen of Program 4.3 after execution.

Here, as shown in Figure 4.3 user is prompted to enter assignment marks of first student. This data is used to print three best marks of assignments. The sum and average of the best assignment marks are also printed. Similarly, user is prompted to enter test marks. The best three test marks are printed. The total and average of the best test marks are also displayed. Similar is the case with data of second and third student.

4.5 Nested Functions

Functions can be called inside another function as a function body. This is called as nesting of functions. Here, we can see how function pointer passes through multiple function call.

General form:

```
function1()
{
    function2();
}
function2()
{
    function3();
}
function3()
{
}
```

Program 4.4: Use of nested functions

Here, three functions are declared. Main function calls function1, function1 in turn calls function2, and function2 calls functions3. Thus, functions are nested.

```
/* Program to demonstrate calling function inside
another function */

#include<stdio.h>
void function1();
void function2();
void function3();
int main()
```

```
{
        function1();
        return 0;
}
void function1()
{
        printf("\n Hello!I am function1 and called
        in main()function\n");
        function2();
}
void function2()
{
        printf("\n Hello! I am function2 and called
        in function1() function\n");
        function3();
}
void function3()
{
        printf("\n Hello! I am function3 and called
        in function2() function\n\n");
}
```

There is no user interaction here. Messages are embedded in the code itself.
Figure 4.4 shows the messages printed by the functions as they are called.

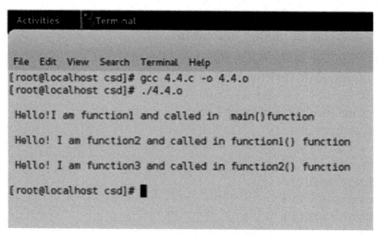

Figure 4.4   Output screen of Program 4.4 after execution.

In the above program, the function1() is called inside the main() function. Here, the main() function is calling function. The function2() is called inside the funtion1() and function3() is called inside function2(). The function2() and function1() work as outer functions, and function3() works as inner function. The sequence of calling function is as defined in main() function. First, main() gets executed, then body of funtion1() sequentially. Program pointer will automatically switch between different function calls.

4.6 Recursion

Define something in terms of itself is the process of recursion. A recursive function is the function whose function body calls the same function. It is also called as "Circular definition". In C, it is possible to call the function in the same function body using recursion.

General form:

funtion1()
{
    function1();
}

Here, function1() in the above syntax is known as recursive function.

Recursion makes code simpler in writing and program more effective. It replaces the lengthy iterations into some recursive statements. But though it looks simple, developing logic to write recursive statement is difficult as compared to writing iterative statements.

Program 4.5: Use of recursive function to add first n natural numbers

Here, value for n is entered by the user and passed to the recursive function. Recursive function will work recursively till the value of n reduces to zero. Here, recursion will break with base case as n=0. When recursion ends, the computed value is returned to main function where it is printed.

```
/* Program to show sum of first n natural numbers
using recursive function */

#include<stdio.h>
int recsum(int n);
```

```
int n1;
int main()
{
        int n,sum;
        sum=0;
        printf("\n Enter number showing first n
        natural numbers...\n");
        scanf("%d",&n);
        sum=recsum(n);
        printf("\n The sum of first %d natural
        numbers is :%d\n\n",n,sum);
        return 0;
}
int recsum(int n)
{
        if(n==0)
                return n;
        else
        {
                n1=n+recsum(n-1);
                return n1;
        }
}
```

User is prompted to enter the value for n, which is entered as 5. The sum of all natural numbers from 1 to 5 is calculated and displayed as 15 as shown in Figure 4.5.

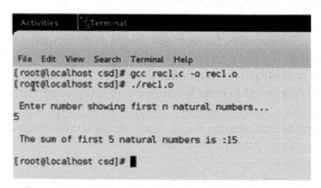

Figure 4.5   Output screen of Program 4.5 after execution.

Program 4.6: Use of recursive function to calculate power of a given number

Here, two parameters are passed to the power function: the number and its power. Recursion breaks when the base condition is met, that is, power = 1.

```c
/* Program to calculate power of a given number
using recursive function */

#include<stdio.h>
int pow1(int n,int p);
int main()

{
        int num,r,ans;
        printf("\n Enter any number\n");
        scanf("%d",&num);
        printf("\n Enter power \n");
        scanf("%d",&r);
        ans=pow1(num,r);
        printf("\n The %d raised to %d is:%d\n\n",
        num,r,ans);
        return 0;
}
int pow1(int n,int p)
{
        int answer;
        if(p!=1)
        {
                answer=n*pow1(n,p-1);
                return answer;
        }

}
```

Here, as shown in Figure 4.6 user is prompted to enter the number and its power. The values entered are 3 and 3, the computed value is 27.

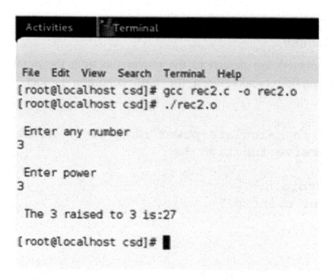

Figure 4.6 Output screen of Program 4.6 after execution.

Program 4.7: Use of recursive function to compute factorial of a number

Using function to calculate factorial of a number is already shown in Program 4.2. The only difference here is that this program function is recursive, whereas in Program 4.2 it is iterative. As it is recursive, the base condition is must to terminate the recursion. Base condition is when n=1.

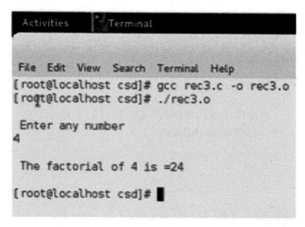

Figure 4.7 Output screen of Program 4.7 after execution.

```
/* Program to calculate factorial of a given number
using recursive function */

#include<stdio.h>
int fact(int n);
int main()
{
        int num,ans;
        printf("\n Enter any number\n");
        scanf("%d",&num);
        ans=fact(num);
        printf("\n The factorial of %d is =%d\n\n",
        num,ans);
        return 0;
}

int fact(int n)

{
        int res;
        if(n!=1)
        {
                res=n*fact(n-1);
                return res;
        }
}
```

User is prompted to enter the number to calculate factorial as shown in Figure 4.7. The entered number is 4, and factorial of 4, that is, 24, is displayed.

Program 4.8: Forming different combinations of binary digits (0, 1) using recursion

Here, no data is accepted from the user. It is embedded in the code itself. This program shows different combinations of two-digit binary numbers using 0 and 1.

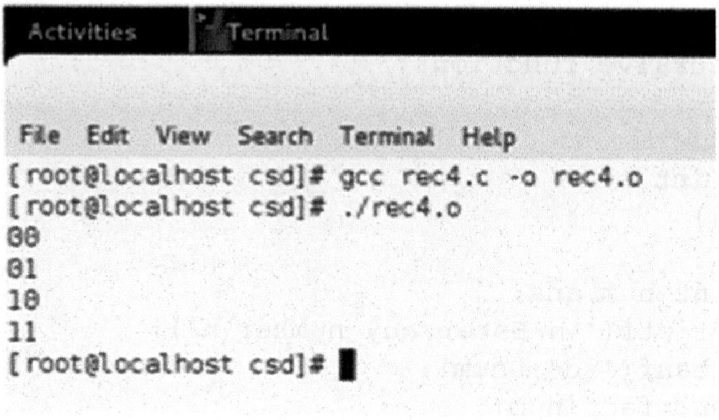

```
Activities        Terminal

File  Edit  View  Search  Terminal  Help
[root@localhost csd]# gcc rec4.c -o rec4.o
[root@localhost csd]# ./rec4.o
00
01
10
11
[root@localhost csd]#
```

Figure 4.8 Output screen of Program 4.8 after execution.

```c
/* Program to show combination of numbers 0 and 1
using recursive function */

#include<stdio.h>
int comb(int n,int cnt);
int main()
{
        comb1(4,2);
        return 0;
}
int comb(int n,int cnt)
{
        if(cnt==0)
        {
                return 0;
        }
        comb(n/2,cnt-1);
        printf("%d",n%2);
}
int comb1(int n,int cnt)
{
        int i;
        for(i=0;i<n;i++)
        {
```

```
comb(i,cnt);
printf("\n");
    }
}
```

As shown in Figure 4.8, Different combinations of 0 and 1 are displayed recursively.

4.7 Conclusion

In this chapter, how to write functions in C was studied. Now, one can define any task in the form of function providing its prototype as given above in declaration and syntax. Each function has function declaration, function definition, and function call statement. Functions save writing code lines again and again. It provides "Write Once, Call Multiple Times" strategy related to program logic statements. There are some functions that return nothing, and can be declared using void data type as their return data type. One can see that in Linux, the return type of main() function is always int, so return statement in the main() function should be added to return positively from the program execution.

4.8 Exercises

1. Write a program for bank transactions using function.
2. Write a program to calculate arithmetic mean of user-entered numbers using function.
3. Write a program to demonstrate nested functions.
4. Write a program to simulate simple calculator using functions.
5. Write a program to calculate mean, mode, and standard deviation using functions.

5

Structure and Dynamic Memory Allocation

5.1 User-Defined Data Types

We use variables of built-in data types to store values in memory in programming constructs. Using built-in data types, we can define a single type of data. The range to store the values is fixed, and we can define any entity using single attribute only. If we have to define any entity with multiple attributes, then we should use array for each attribute. It is possible to use multiple arrays in a program to define any entity, but it is not the proper as well as easy way to write the program code. It increases program length and complexity also. So, C helps us to define these entities using structure. Using structure, we can define entities such as book, person, table, and pen.

For example, person has attributes like name, age, height, and gender.

Before structure, if the problem is to display information about 10 people then you should declare separate arrays for name, age, height and gender etc. If the attributes go on increasing, writing program with this technique becomes a complex task. So there is a need for structures. We need to write person structure as follows.

structPerson

```
{

        int age;
        char name[20];
        float height;
        char gender;

};
```

5.2 Defining Structure

Before using structures in the program they should be declared indicating the types of members embedded in them. The syntax for declaration is given as follows:

Syntax:
structstructure_name
{
    data_type variable;
    data_type variable;
};

Here, struct is the keyword used to declare a structure.

    structure_name is the user-defined name given to the structure we are going to define.

    data_type is the valid C data type.

    variable is the named memory location which is used to store values.

Consider the above example here.

structPerson
{
    int age;
    char name[20];
    float height;
    char gender;
};

Here, the above statement defines a new user-defined data type called **structPerson**. The following program demonstrates the use of simple structure to store and display date.

Program 5.1: Program to show the use of structure to store and display date

Date consists of three parts, namely day, month, and year. So here, structure is created with the name date having members as day which will accept day of the month and is an integer value. The second member of the structure is month which is also an integer variable between 1 and 12. The last member of the structure is year which again is an integer variable. It is a simple program without any validations.

```c
/* Program to implement the date structure */

#include<stdio.h>
int main()
{
        struct date
        {
                int day;
                int month;
                int year;

        };
        struct date d1;
        printf("\n Enter date as day , month and
        year:\n");
        scanf("%d%d%d",&d1.day,&d1.month,&d1.year);
        printf("\n Entered Date is: %d/%d/%d\n\n",
        d1.day,d1.month,d1.year);
        return 0;
}
```

As shown in Figure 5.1, the user is prompted to enter day, month, and year. The entered value is displayed in the date format.

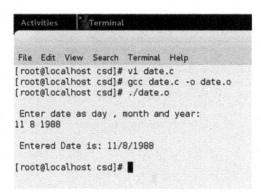

Figure 5.1   Output screen of Program 5.1 after execution.

Program 5.2: Program to show the use of structure to store and display time

It is similar to Program 5.1; instead of day, month, and year, the user has to give input as hour, minute, and seconds.

```c
/* Program to implement the time structure */

#include<stdio.h>
int main()
{

        struct time
        {
                int hour;
                int minute;
                int second;
        };
        struct time t1;
        printf("\n Enter time as hour, minute and
        second:\n");
        scanf("\n%d%d%d",&t1.hour,&t1.minute,&t1
        .second);
        printf("\nEntered time is:: %d:%d:%d\n\n",
        t1.hour,t1.minute,t1.second);
        return 0;
}
```

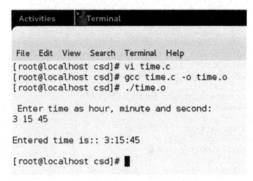

Figure 5.2   Output of Program 5.2 after execution.

As shown in Figure 5.2, the user is prompted to enter hour, minute, and second values in the integer form. The same is displayed in the time format. There is no validation carried out here.

Program 5.3: Program to show the use of array of structure

This is a very good example showing the use of structure to store heterogeneous data, viz. character, integer, and float. Compared to array, this is an advantage of structure as arrays are homogeneous in nature and cannot store different data types under one name.

Here, a structure with the name student is created to store the data related to student like name, roll number, and percentage of marks. Then, this structure is declared as an array which can hold data of 10 students.

```
/* Program to read and display the student informa-
tion using structure */

#include<stdio.h>
int main()
{
        struct student
        {
                char nm[10];
                int rno;
                float per;
        };
        inti,no;
        struct student s[10];
        printf("\n Enter number of students for
        whom you want to enter details: \n");
        scanf("%d",&no);
        if(no<=10)
        {
                for(i=0;i<no;i++)
                {
                        printf("\n Enter student
                        details: Name,Roll Number
                        and Marks in percentage:\n");
                        scanf("%s%d%f", &s[i].nm,
                        &s[i].rno,&s[i].per);
```

```
            }
            printf("\n");
    printf("\n Student Information is:\n");
    printf("----------------------------\n");
    for(i=0;i<no;i++)
    {
    printf("%s \t %d \t %.2f \n",s[i].nm,s[i]
    .rno, s[i].per);
    }
    }
    else
    {
            printf("\n The number of students
            should be 10 or less than 10\n");
    }
    return 0;
}
```

In Figure 5.3, it can be observed that the user is prompted to enter the number of students whose data needs to be stored. As embedded in the code, this number should be less than or equal to 10; otherwise, a message is

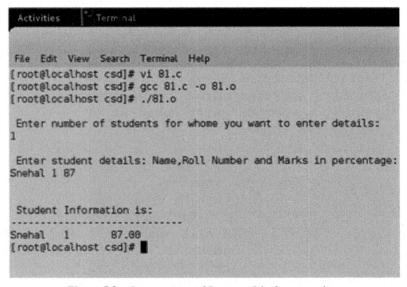

Figure 5.3 Output screen of Program 5.3 after execution.

displayed to enter the number less than or equal to 10. Here, the user enters one and the data for single student is accepted and displayed as shown in Figure 5.3.

Program 5.4: Program to show the use of structure for storing and searching data

This program structure is same as Program 5.3. It has an extra for loop to search for a particular roll number entered by the user. If found, then display the details of the same or else prompt the user to enter a valid roll number.

```c
/* Program to display the student information of
given roll number using structure */

#include<stdio.h>
struct student
{
        int rno;
        char nm[10];
        float per;
};
int main()
{
        struct student s[10];
        inti,f,no,roll;
        f=0;
        printf("\n Enter number of students:\n");
        scanf("%d",&no);
        for(i=0;i<no;i++)
        {
                printf("\n Enter student information
                as Name,Roll Number and Marks in
                Percentage\n");
                scanf("%s\t%d\t%f",&s[i].nm,&s[i]
                .rno,&s[i].per);
        }
printf("\n Student Information:\n");
        printf("\n Enter Roll Number of a student
        \n");
```

```
scanf("%d",&roll);
for(i=0;i<no;i++)
{
        if(roll==s[i].rno)
        {
                f=1;
                break;
                break;
        }
}
if(f==1)
        printf("\n Roll No=%d\t Name of the
        Student=%s\tMarks in Percentage=
        %.2f",s[i].rno,s[i].nm,s[i].per);
else
        printf("\n Enter proper roll
        number\n");
printf("\n\n");
}
```

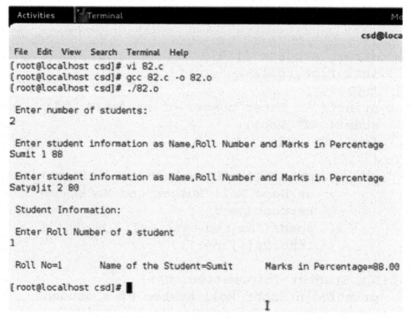

Figure 5.4   Output screen of Program 5.4 after execution.

Here, as shown in Figure 5.4 the user is prompted to enter the number of students whose data needs to be stored. The user has entered 2. Data related to two students is stored, and the user is again prompted to enter the roll number whose details need to be searched. It is entered as 1, which is a valid roll number, so its details are displayed.

Program 5.5: Program to show the use of structure to analyze students' marks stored

Again, the student structure remains the same as in Program 5.4, but here instead of searching for the roll number, the details of students are displayed who have scored greater than or equal to 80%.

```
/* Program to display the student information
who have secured marks greater than or equal to
80 percent */

#include<stdio.h>
int main()
{
struct student
    {
char nm[10];
intrno;
float per;
    };
inti,no;
struct student s[10];
printf("\n Enter number of students for whom you
want to enter details: \n");
scanf("%d",&no);
if(no<=10)
    {
for(i=0;i<no;i++)
        {
printf("\n Enter student details: Name,Roll Number
and Marks in percentage:\n");
scanf("%s%d%f",&s[i].nm,&s[i].rno,&s[i].per);
        }
```

```
printf("\n Students details having
marks greater than 80%\n");
for(i=0;i<no;i++)
{
        if(s[i].per>=80)
        {
                printf("\n Name=%s\
                tRoll Number=%d\
                tPercentage=%.2f\n",
                s[i].nm,s[i].rno,s[i]
                .per);
        }
}
}
else
{
        printf("\n The number should be 10
        or less than 10");
}
printf("\n\n");
}
```

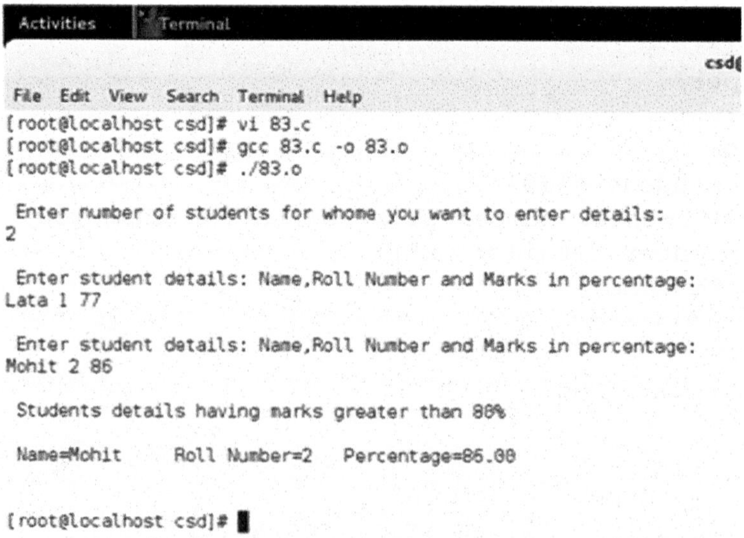

Figure 5.5   Output screen of Program 5.5 after execution.

Here, as shown in Figure 5.5 data of students is accepted as entered by the user. This data is then analyzed, and the details of the students with percentage greater than or equal to 80% are displayed.

Program 5.6: Program to show the use of function with structure as parameter

Here, the employee details are stored, and then, the salary variable is analyzed to increment. If the salary is less than or equal to 2000, then 15% increment is given, if the salary is greater than 2000 and less than or equal to 5000, then increment of 10% is given, and if the salary is greater than 5000, then there is no increment.

```
/* Program to display the employee information and
increase his/her salary as per the given criteria
using structure and function */

#include<stdio.h>
int display(int);
int main()
{

        struct employee
        {
        int eno, day, yr;
        char nm[10], month[10];
        intsal;
        };
        int i,no;
        struct employee e[10];
        printf("\n Enter number of employee\n");
        scanf("%d",&no);
        if(no<=10)
        {
                for(i=0;i<no;i++)
                {
                    printf("\n Enter Employee details
                    as EmpNumber,Name,Salary and Date
                    of joining as Day,Month,Year\n");
```

```
                scanf("%d%s%d%d%s%d",&e[i].eno,
                &e[i].nm,&e[i].sal,&e[i].day,
                &e[i].month,&e[i].yr);
            }
            for(i=0;i<no;i++)
            {
                display(e[i].sal);
            }
        }
        else
            printf("\n The number of employee
            should be 10 or less than 10");
        printf("\n\n");
}
int display(int sal1)
{
        if(sal1<=2000)

        {
                sal1=sal1+((sal1*15)/100);
                printf("\n The salary is increased by
                15%=%d",sal1);

        }
        else if((sal1<=5000)&&(sal1>2000))
        {
                sal1=sal1+((sal1*10)/100);
                printf("\n The salary is increased by
                10%=%ld",sal1);
        }
        else if(sal1>5000)
        {
                printf("\n Salary will not be
                incremented \n",sal1);
        }
        else
                printf("\n Enter valid Salary\n");


}
```

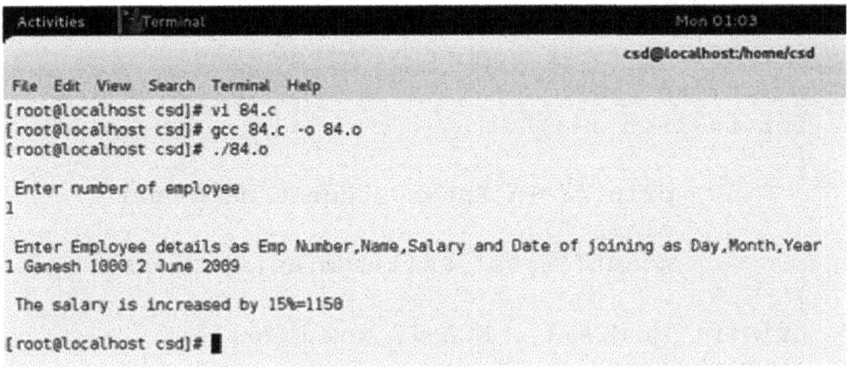

Figure 5.6 Output screen of Program 5.6 after execution.

The user is prompted to enter the number of employees, and accordingly, the data related to employees is accepted from the user as shown in Figure 5.6. As mentioned above, the increment in the salary is calculated and displayed.

Program 5.7: Program to display the students' details classwise

Here, the user enters the student name as well as class name. After accepting all the students' data, it prints the details of students classwise. Likewise, if class is MCA-I, then all students' details from this class will be displayed under the label MCA-I. The program uses two character arrays as members of structure student to accept the name and class name.

```c
/* Program to display students classwise using
structure */

#include<stdio.h>
#include<string.h>
int main()
{
      struct student
      {
            char nm[10],class[10];
      }
      s[5];
```

```
int i,j,n;
printf("\n Enter number of students\n");
scanf("%d",&n);
for(i=0;i<n;i++)
{
        printf("\n Enter student name and
        class:");
        scanf("%s%s",&s[i].nm,&s[i].class);
}
printf("\n MCA-I , MCA-II and MCA_III
students");
printf("\n");
printf("\n MCA-I Students");
for(i=0;i<n;i++)
{
        j=strcmp(s[i].class,"MCA-I");
        if(j==0)
                printf("\n%s\t%s",s[i].nm,
                s[i].class);
}
printf("\n MCA-II Students");
for(i=0;i<n;i++)
{
        j=strcmp(s[i].class,"MCA-II");
        if(j==0)
                printf("\n%s\t%s",s[i].nm,
                s[i].class);
}
printf("\n MCA_III Students");
for(i=0;i<n;i++)
{
        j=strcmp(s[i].class,"MCA-III");
        if(j==0)
                printf("\n%s\t%s",s[i].nm,
                 s[i].class);
}
printf("\n\n");
}
```

Figure 5.7 Output screen of Program 5.7 after execution.

The user is prompted to enter the number of students whose data needs to be accepted. Here, the details of five students are accepted. Then, the list is displayed classwise as shown in Figure 5.7.

5.3 Nesting of Structure

Structures can be nested within one another. We can use one structure variable as a data member of another structure. When one structure variable can be used as another structure member, then it is called as nested structure. It is useful when one structure definition has to be used more than once in a program. We can simply create number of structure variables rather than defining it again and again. Example: **employee** structure, the date of birth and date of joining can be structure variables of **date** structure and data members of **employee** structure.

The following program shows that how structures can be nested within one another.

Program 5.8: Program to show the use of structure within structure

Here, in this program, there are two structures. The first structure is date, and the second is person. Person structure has date of birth as member. Date of birth consists of three items, i.e., day, month, and year. So date is declared as structure and is made member of Person structure. Person structure has the members name, age, and date of birth.

```c
/* Program to demonstrate the nested structures */

#include<stdio.h>
int main()
{
    struct date
    {
        int day;
        int month;
        int year;
    };
    struct person
    {
        char name[15];
        nit age;
        struct date d1;
    };
    struct person p1;
    printf("\n Enter your name, age and birth
    date of a person\n");
    scanf("\n%s%d%d%d%d",&p1.name,&p1.age,&p1.d1
    .day,&p1.d1.month,&p1.d1.year);
    printf("\n Your Information\n");
    printf("\nName: %s\nAge: %d\nBirth Date:
    %d/%d/%d\n\n",p1.name,p1.age,p1.d1.day,p1.d1
    .month,p1.d1.year);
    return 0;
}
```

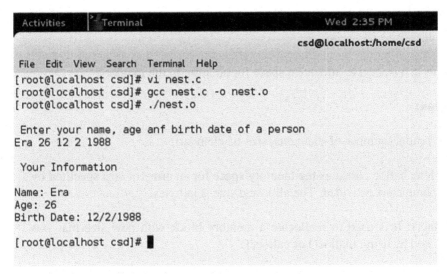

Figure 5.8 Output screen of Program 5.8 after execution.

As shown in Figure 5.8, the user is prompted to enter his details, and the same is printed in a proper format.

5.4 Dynamic Memory Allocation

Dynamic memory allocation is a unique feature of C language. It allows us to create data types and data structures of any size and length that we need in our program. In dynamic allocation of memory, certain amount of memory is requested from the memory heap. If request is granted, then that much amount of memory is reserved for the program. When you finish with program then you should return it to the operating system.

The C language provides malloc(), calloc(), realloc(), and free() functions to achieve these strength of C. These functions are available in stdlib.h header file.

malloc(): It is used to allocate a memory block that is being used in our program.

Syntax:

void *malloc(size);

Here, malloc() returns pointer to a block of memory of size specified by *size* parameter. The allocated size is in bytes. If the requested amount of memory is not available, then malloc() returns NULL. If return value is

NULL, then the program stops. The return values of malloc() are either void or null.

calloc(): It is used to allocate a space for an array in the memory.

Syntax:

void *calloc(number-of-elements, size-of-element);

Here, calloc allocates the memory space for an array of size specified by two parameters provided. The allocated size is in bytes.

realloc(): It is used to reallocate a memory block with new size that was allocated by using malloc() or calloc().

Syntax:

void *realloc(pointer,size);

Here, realloc() changes the size of memory allocated previously pointed by the *pointer* parameter and allocates new size specified by *size* parameter. The allocated memory is in bytes.

free(): It is used to deallocate memory that was allocated by using malloc() or calloc().

Syntax:

void free(pointer);

Here, free() deallocates memory block previously allocated and pointed by the *pointer* parameter.

Program 5.9: Program to show the use of malloc function for dynamic memory

Dynamic memory allocation is required whenever there is a pointer variable in the program. Here, dynamic array is created using pointer. The size of the array is accepted from the user at run time. According to the size entered by the user, the array is declared and elements are added. The sum and average of these elements are computed and displayed.

```c
/* Program to demonstrate the use of dynamic memory
allocation to find the sum and average of array
elements */

#include<stdio.h>
#include<stdlib.h>
int main()
{
    int *arr,n,i,sum;
    sum=0;
    floatavg;
    printf("\n Enter number of elements:\n");
    scanf("%d",&n);
    arr=(int *)malloc(n*sizeof(int));
    printf("\n Enter values for the elements:\n");
    for(i=0;i<n;i++)
    {
        scanf("%d",&arr[i]);
        sum=sum+arr[i];
    }
    avg=sum/n;
    printf("\n Sum=%d\n Average=%.2f",sum,avg);
    printf("\n\n");
    return 0;
}
```

As shown in Figure 5.9, the user is prompted to enter the size of the array, which is entered as 5. Then, the array elements are accepted from the user and stored. The sum and average of these elements are displayed.

Program 5.10: Program to show the use of dynamic structure

Here, a pointer of type student structure is declared to point to the structure. Memory to this structure is allocated dynamically using malloc and calloc functions. Student details are accepted, stored, and displayed.

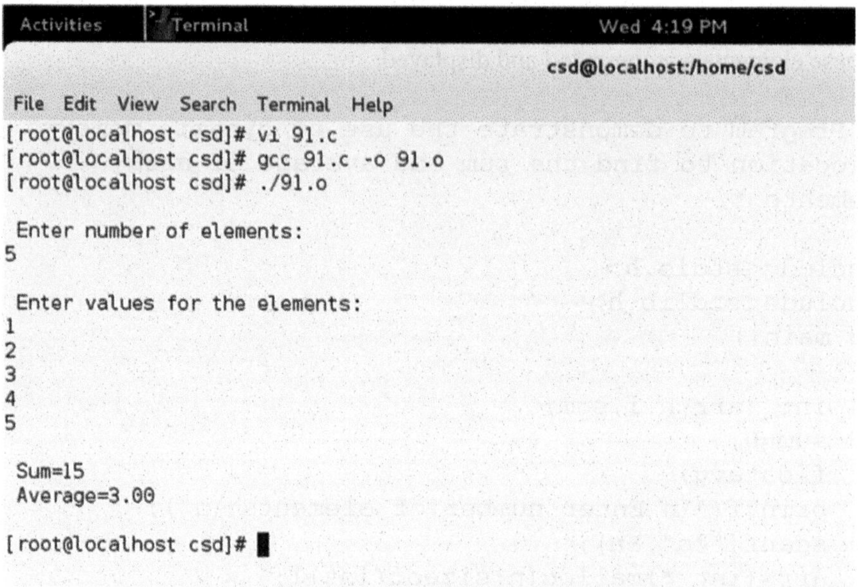

```
Activities          Terminal                              Wed 4:19 PM

                                              csd@localhost:/home/csd

File  Edit  View  Search  Terminal  Help
[root@localhost csd]# vi 91.c
[root@localhost csd]# gcc 91.c -o 91.o
[root@localhost csd]# ./91.o

Enter number of elements:
5

Enter values for the elements:
1
2
3
4
5

Sum=15
Average=3.00

[root@localhost csd]# █
```

Figure 5.9   Output screen of Program 5.9 after execution.


```c
/* Program to demonstrate the dynamic memory
allocation to display the student information
using structure */

#include<stdio.h>
#include<stdlib.h>
int main()
{
        struct student
        {
                intrno;
                char nm[10],addrs[20];
        };
        inti,n;
        struct student *s;
        s=(struct student *)malloc(1*sizeof(struct
        student));
        s=(struct student *)calloc(2,2);
        printf("\n Enter number of students:\n");
```

```
        scanf("%d",&n);
        s=realloc(s,n);
for(i=0;i<n;i++)
        {
                printf("\n Enter student
information as roll number,name and address");
                scanf("%d%s%s",&s[i].rno,&s[i].nm,
                &s[i].addrs);
        }
        for(i=0;i<n;i++)
        {
                printf("\n Details of Students are:
                \n");
                printf("Roll Number=%d\t Name=%s\t
                Address=%s",s[i].rno,s[i].nm,s[i]
                .addrs);
        }
        printf("\n\n");
        return 0;
}
```

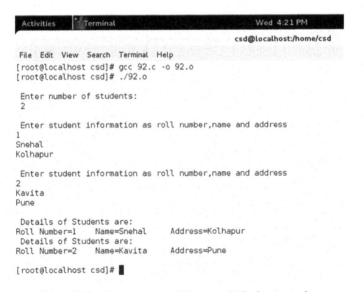

Figure 5.10   Output screen of Program 5.10 after execution.

The user is prompted to enter the number of students, which is entered as two. Then, data related to two students is accepted and displayed as shown in Figure 5.10.

5.5 Conclusion

In this chapter, we have studied structure as a user-defined data type. Now, we can define any user-defined data type using structure. For example, student, person, class, book etc. also, it is possible to use one structure within another structure called as nested structures. Also, we have studied the dynamic memory allocation in C which is the C's most powerful feature. Malloc() returns void pointer and it can be converted to another data type to store data of built-in as well as user-defined data type.

5.6 Exercises

1. Write a program to define a Book structure and show book details.
2. Write a program to define a structure which accepts valid date.
3. Write a program to store 5 products using Product structure with data members as id, name, price, expiry_date, and manufacturing_date.
4. Write a program to define Student structure. Display student's details along with his/her department details. Define department structure to show department details.
5. Write a program to define a structure which displays one's personal and academic profile.

6

Data Structures

6.1 Data Structure

Data structure is a mathematical or logical organization of data in memory that consider not only the items stored but also relationship between the items stored. This relationship between data items allows the designing of efficient algorithms for manipulation of data items. There are two categories of data structures: linear and nonlinear. Linear data structures define arrangement of data items in a linear fashion. Also, we access and manipulate those data items in a linear fashion. Nonlinear data structures are not in the scope of this book. We have seen one of the linear data structure Array to store and access data items of similar type. Data structure shows relationship between data items, and it depends upon which data structure that is in use.

The following are linear data structures:

- Array
- Stack
- Queue
- Linked List.

6.2 Stack

When we talk about stack as a data structure, then there are two things associated with it: first is its features or characteristics and second is the set of operations associated with it to manipulate its data.

6.2.1 Features of Stack

Stack is collection of similar data types, with access at only one end. Following features describes stack as a data structure.

- Stack is a linear data structure used to store similar type of data items.
- In stack, data items are inserted and accessed by one end only.
- It is also called as Last in First out (LIFO) data structure.
- It is a unidirectional data structure.
- Stack can be implemented in two ways:
    - Static implementation: Here, Stack is implemented using arrays.
    - Dynamic implementation: Here, Stack is implemented using Linked structure.

6.2.2 Operations on Stack

PUSH: This operation inserts a new data item into the stack. Initially, the stack pointer is at 0th location. Each time the data item is inserted, the stack pointer is incremented by one. The last data item in the stack to which stack pointer is pointing is called as top of the stack. Figure 6.1 shows the operations of stack.

POP: This operation deletes a data item from the stack. Each time you perform this operation, the data item presented at the top of the stack is deleted. Every delete operation decreases the stack pointer by one.

Traverse: This operation visits all the data items stored in the stack one by one and is mainly used to print contents of the stack.

Return Top: This operation returns the top data item of the stack without deleting the data item from the stack.

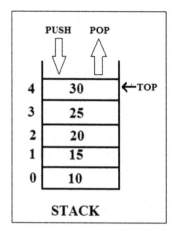

Figure 6.1 Operations of stack.

Program 6.1: Implementation of stack using array

Here, static stack is declared using array As already discussed, stack is a Last In First Out (LIFO) data structure with the operations PUSH to insert the item, POP to delete the item, and Traverse to display the items in the stack.

It is a menu-driven program asking user to enter the operation they want to perform on stack.

```c
/* Program to show static implementation of stack */

#include<stdio.h>
#define MAX 5
int top = -1;
intstack_arr[MAX];
void main()
{
        int choice;
        while(1)
        {
                printf("1.Push\n");
                printf("2.Pop\n");
                printf("3.Display\n");
                printf("4.Quit\n");
                printf("Enter your choice : ");
                scanf("%d",&choice);
                switch(choice)
                {
                case 1:
                    push();
                    break;
                case 2:
                    pop();
                    break;
                case 3:
                    display();
                    break;
                case 4:
                    return 0;
```

```c
                        default:
                                printf("Wrong choice\n");
                                break;
                }
        }
}
push()
{
        intpushed_item;
        if(top == (MAX-1))
                printf("Stack Overflow\n");

        else

        {
                printf("Enter the item to be pushed
                in stack : ");
                scanf("%d",&pushed_item);
                top=top+1;
                stack_arr[top] = pushed_item;
        }
}
pop()
{
        if(top == -1)
                printf("Stack Underflow\n");
        else
        {
                printf("Popped element is : %d\n",
                stack_arr[top]);
                top=top-1;
        }
}
display()
{
        inti;
        if(top == -1)
                printf("Stack is empty\n");
        else
```

```
{
        printf("Stack elements :\n");
        for(i = top; i>=0; i--)
            printf("%d\n", stack_arr[i] );
}
}
```

As shown in Figure 6.2, user is prompted to select a menu item from the menu, and according to the selection, operations on stack are carried out. Quit menu item is selected to stop the program.

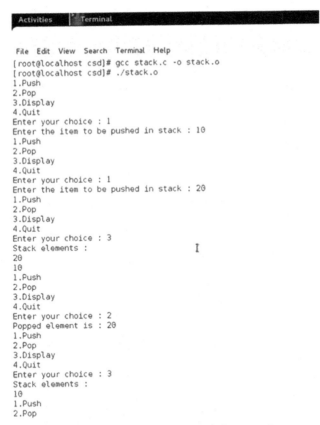

Figure 6.2 Output of Program 6.1 after execution.

Program 6.2: Use of structure in implementing stack

Here, pointer is used in implementation. So as compared to Program 6.1, stack size is not fixed here. When item is pushed, memory is allocated, and when the item is popped, memory is released, thus making it memory-efficient implementation.

```c
/* Program to show dynamic implementation
of stack */

# include<stdio.h>
# include<malloc.h>
struct node
{
        int info;
        struct node *link;
}
*top=NULL;
main()
{
        int choice;
        while(1)
        {
            printf("1.Push\n");
            printf("2.Pop\n");
            printf("3.Display\n");
            printf("4.Quit\n");
            printf("Enter your choice : ") ;
            scanf("%d", &choice);
            switch(choice)
            {
            case 1:
                    push();
                    break;
            case 2:
                    pop();
                    break;
            case 3:
                    display();
```

```
                        break;
                case 4:
                        return 0;
                default:
                        printf("Wrong choice\n");
                        break;
                }
        }
}
push()
{
        struct node *tmp;
        int pushed_item;
        tmp = (struct node *)malloc(sizeof(struct
        node));
        printf("Input the new value to be pushed
        on the stack : ");
        scanf("%d",&pushed_item);
        tmp->info=pushed_item;
        tmp->link=top;
        top=tmp;
}
pop()
{
        struct node *tmp;
        if(top == NULL)
            printf("Stack is empty\n");
        else
        {
           tmp=top;
           printf("Popped item is %d\n",tmp->info);
           top=top->link;
           free(tmp);
        }
}
display()
{
        struct node *ptr;
        ptr=top;
```

```
if(top==NULL)
    printf("Stack is empty\n");
else
{
    printf("Stack elements :\n");
    while(ptr!= NULL)
    {
            printf("%d\n",ptr->info);
            ptr = ptr->link;
    }
}
}
```

The output shown in Figure 6.3 is almost the same as shown in Figure 6.2 except that implementation is dynamic, not static.

Figure 6.3   Output of Program 6.2 after execution.

6.3 Queue

Queue is a linear data structure in which data is inserted at one end called as REAR and data is deleted from another end called as FRONT.

6.3.1 Features of Queue

Queue is a homogeneous data structure as it has:

- It is a linear data structure used to store similar type of data items.
- In Queue, data items are inserted at one end and deleted from other end. It maintains two access points, Front and Rear.
- It is also called as First In First Out (FIFO) data structure.
- It is unidirectional in nature.
- Queue can be implemented in two ways:
  - Static implementation: Here, Queue is implemented using arrays.
  - Dynamic implementation: Here, Queue is implemented using Linked structure.

6.3.2 Operations on Queue

Queue performs following operations on data items. Figure 6.4 shows the characteristic of queue data structure.

INSERT: This operation inserts a new data item to the queue. Each time data item is inserted to the queue, REAR is incremented by one.

DELETE: This operation deletes a data item from the queue. Each time data item is deleted from the queue, the FRONT is incremented by one.

Traverse: This operation visits all the data items stored in the Queue one by one and is mainly used to print contents of the Queue.

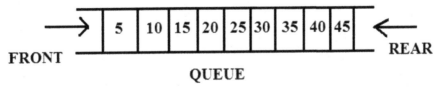

Figure 6.4   Data structure queue.

Program 6.3: Implementation of queue using array

Here, queue is declared as int array, and index variables as rear and front for insertion and deletion. This program is menu-driven and also takes care of overflow and underflow conditions.

```c
/* Program to show implementation of Queue */

#include<stdio.h>
#define MAX 5

int queue_arr[MAX];
int rear = -1;
int front = -1;
void main()
{
        int choice;
        while(1)
        {
                printf("1.Insert\n");
                printf("2.Delete\n");
                printf("3.Display\n");
                printf("4.Quit\n");
                printf("Enter your choice : ");
                scanf("%d",&choice);
                switch(choice)
                {
                case 1:
                        insert();
                        break;
                case 2:
                        del();
                        break;
                case 3:
                        display();
                        break;
                case 4:
                        return 0;
                default:
```

```
                        printf("Wrong choice\n");
                        break;
                }
        }
}
insert()
{
        intadded_item;
        if (rear==MAX-1)
                printf("Queue Overflow\n");
        else
        {
                if (front==-1)
                        front=0;
                printf("Input the element for adding
                in queue : ");
                scanf("%d", &added_item);
                rear=rear+1;
                queue_arr[rear] = added_item ;
        }
}
del()
{
        if (front == -1 ||front > rear)
        {
                printf("Queue Underflow\n");
                return ;
        }
        else
        {
                printf("Element deleted from queue
                is : %d\n", queue_arr[front]);
                front=front+1;
        }
}
display()
{
        inti;
```

```
    if (front == -1)
            printf("Queue is empty\n");
    else
    {
            printf("Queue is :\n");
            for(i=front;i<= rear;i++)
                    printf("%d ",queue_arr[i]);
            printf("\n");
    }
}
```

User is prompted to enter his choice from the menu being displayed. According to the choice, the operations are carried out on queue and displayed as shown in Figure 6.5.

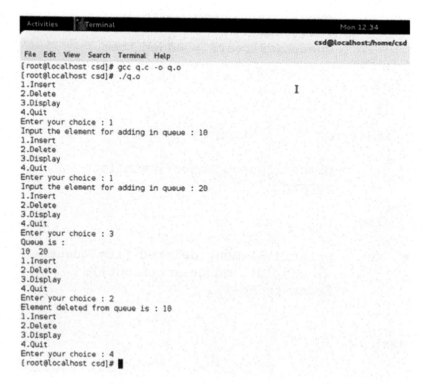

Figure 6.5 Output of Program 6.3 after execution.

Program 6.4: Use of pointer to structure to implement dynamic queue

Here, memory allocation is not static as in Program 6.3. Memory is allocated dynamically using malloc function; hence, chances of queue overflow are reduced.

```c
/* Program to implement Dynamic implementation
of queue */
# include<stdio.h>
# include<malloc.h>
struct node
{
        int info;
        struct node *link;
}
*front=NULL,*rear=NULL;
void main()
{
        int choice;
        while(1)
        {
                printf("1.Insert\n");
                printf("2.Delete\n");
                printf("3.Display\n");
                printf("4.Quit\n");
                printf("Enter your choice : ");
                scanf("%d", &choice);
                switch(choice)
                {
                case 1:
                        insert();
                        break;
                case 2:
                        del();
                        break;
                case 3:
                        display();
                        break;
                case 4:
                        return 0;
```

```
                default:
                        printf("Wrong choice\n");
                        break;
                }
        }
}
insert()
{
        struct node *tmp;
        intadded_item;
        tmp = (struct node *)malloc(sizeof(struct
        node));
        printf("Input the element for adding in
        queue : ");
        scanf("%d",&added_item);
        tmp->info = added_item;
        tmp->link=NULL;
        if(front==NULL)
                front=tmp;
        else
                rear->link=tmp;
        rear=tmp;
}
del()
{
        struct node *tmp;
        if(front == NULL)
                printf("Queue Underflow\n");
        else
        {
                tmp=front;
                printf("Deleted element is %d\n",
        tmp->info);
                front=front->link;
                free(tmp);
        }
}
display()
{
```

```
struct node *ptr;
ptr = front;
if(front == NULL)
        printf("Queue is empty\n");
else
{
        printf("Queue elements :\n");
        while(ptr != NULL)
        {
                printf("%d ",ptr->info);
                ptr = ptr->link;
        }
printf("\n");
        }
}
```

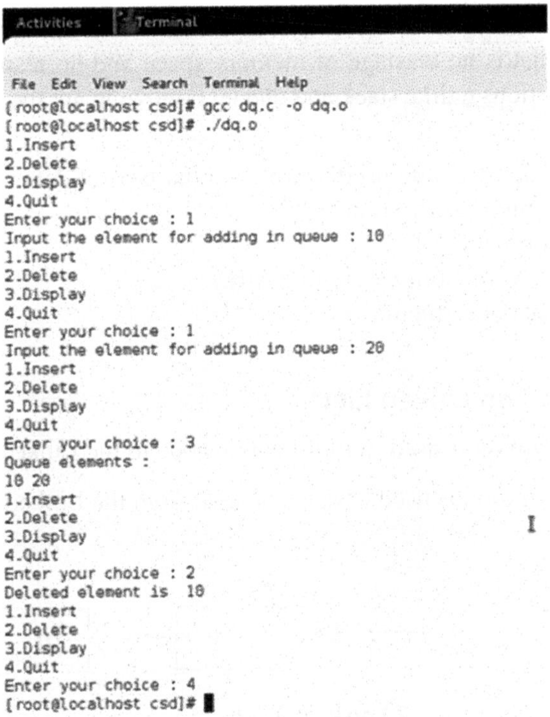

Figure 6.6 Output of Program 6.4 after execution.

As shown in Figure 6.6, user is prompted to enter the menu choice. Operations are carried out on queue depending on the user choice. Results are displayed on the screen of operations.

6.4 Linked List

In Liked List data structure, one can create node consisting of two parts, data part and address part. Data part defines the actual value of the data item, and address part defines address of next node to which current node is connected. There is no any constrain on Linked List data structure. One can insert node before or after to the current node. Linked List can be used as base for dynamic implementation of data structure such as Stack, Queue, Tree, and Graph. It is a basic dynamic data structure.

6.4.1 Features of Linked List Pictorial Representation of Linked List Is Shown in Figure 6.7

It is a linked structure more flexible as compared to stack and queue. Main feature of linked list is no wastage of memory space and no restrictions on insertions and deletions unlike stack and queue. Following are the features of linked list.

- It is a linear data structure used to store similar type of data items.
- It allows arbitrary insertions and deletions.
- It is unidirectional in nature.
- It does not allow random access like Array.
- It is a dynamic data structure.

6.4.2 Operations on Linked List

INSERT: This operation is used to insert a new node to the Linked List.

DELETE: This operation is used to delete a node from the Linked List.

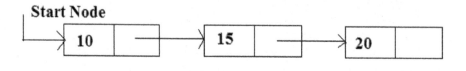

Linked List

Figure 6.7   Data structure Linked List.

Traverse: This operation is used to visit each node of Linked List once.

Searching & Sorting: This operation is used to search for particular data item and sort the data items in the Linked List.

Program 6.5: Implementation of dynamic data structure

Linked list is a dynamic data structure. Here, it is implemented using pointers to the structure. Many operations are presented here such as insert after and at the beginning, delete an item from the linked list, reverse the contents of linked list, search an item in linked list, and traverse link list.

```c
/* Program to show implementation of Linked List */

#include<stdio.h>
#include<malloc.h>
struct node
{

        int info;
        struct node *link;
}

*start;
void main()
{

        intchoice,n,m,position,i;
        start=NULL;
        while(1)
        {

                printf("1.Create List\n");
                printf("2.Add at beginning\n");
                printf("3.Add after \n");
                printf("4.Delete\n");
                printf("5.Display\n");
                printf("6.Count\n");
                printf("7.Reverse\n");
                printf("8.Search\n");
```

```
printf("9.Quit\n");
printf("Enter your choice : ");
scanf("%d",&choice);
switch(choice)
{
case 1:
        printf("How many nodes you
        want: ");
        scanf("%d",&n);
        for(i=0;i<n;i++)
        {
            printf("Enter the
        element: ");
            scanf("%d",&m);
            create_list(m);
        }
        break;
case 2:
        printf("Enter the element : ");
        scanf("%d",&m);
        addatbeg(m);
        break;
case 3:
        printf("Enter the element : ");
        scanf("%d",&m);
        printf("Enter the position
        after which this element is
        inserted :");
        scanf("%d",&position);
        addafter(m,position);
        break;
case 4:
        if(start==NULL)
        {
            printf("List is empty\n");
            continue;
        }
        printf("Enter the element for
        deletion : ");
```

```
                scanf("%d",&m);
                del(m);
                break;

        case 5:
                display();
                break;
        case 6:
                count();
                break;
        case 7:
                rev();
                printf("Linked list after
                reversing is :\n");
                display1();
                break;
        case 8:
                printf("Enter the element
                to be searched : ");
                scanf("%d",&m);
                search(m);
                break;
        case 9:
                return 0;
        default:
                printf("Wrong choice\n");
                break;
                }

        }
}

create_list(int data)

{
        struct node *q,*tmp;
        tmp= malloc(sizeof(struct node));
        tmp->info=data;
        tmp->link=NULL;
        if(start==NULL)
```

```
                start=tmp;
        else
        {
                q=start;
                while(q->link!=NULL)
                q=q->link;
                q->link=tmp;
        }
}

addatbeg(int data)

{

        struct node *tmp;
        tmp=malloc(sizeof(struct node));
        tmp->info=data;
        tmp->link=start;
        start=tmp;
}

addafter(intdata,intpos)

{

        struct node *tmp,*q;
        inti;
        q=start;
        for(i=0;i<pos-1;i++)
        {
                q=q->link;
                if(q==NULL)
                {
                        printf("There are less than %d
                        elements",pos);
                        return;
                }
        }
        tmp=malloc(sizeof(struct node) );
        tmp->link=q->link;
        tmp->info=data;
```

```
            q->link=tmp;

}

del(int data)
{
        struct node *tmp,*q;
        if(start->info == data)
        {
                tmp=start;
                start=start->link;
                free(tmp);
                return;
        }
        q=start;
        while(q->link->link != NULL)
        {
                if(q->link->info==data)
                {
                q->link=tmp->link;
                        tmp=q->link;
                        free(tmp);
                        return;
                }
                q=q->link;
        }
        if(q->link->info==data)
        {
                tmp=q->link;
                free(tmp);
                q->link=NULL;
                return;
        }
        printf("Element %d not found\n",data);
}
display()
{
        struct node *q;
        if(start == NULL)
```

```
        {
                printf("List is empty\n");
                return;
        }
        q=start;
        printf("List is :\n");
        while(q!=NULL)
        {
                printf("%d ", q->info);
                q=q->link;
                }
                printf("\n");
}
count()
{
        struct node *q=start;
        intcnt=0;
        while(q!=NULL)
        {
                q=q->link;
                cnt++;
        }
        printf("Number of elements are %d\n",cnt);
}
rev()
{
        struct node *p1,*p2,*p3;
        if(start->link==NULL)
                return;
p1=start;
        p2=p1->link;
        p3=p2->link;
        p1->link=NULL;
        p2->link=p1;
        while(p3!=NULL)
        {
                p1=p2;
                p2=p3;
                p3=p3->link;
```

```
                p2->link=p1;
        }
        start=p2;
}
search(int data)
{
        struct node *ptr = start;
        intpos = 1;
        while(ptr!=NULL)
        {
                if(ptr->info==data)
                {
                        printf("Item %d found at
                        position%d\n", data,pos);
                        return;
                }
                ptr = ptr->link;
                pos++;
        }
        if(ptr == NULL)
                printf("Item %d not found in list\n",
                data);
}
display1()
{
        struct node *q;
        if(start == NULL)
        {
                printf("List is empty\n");
                return;
        }
        q=start;
        while(q!=NULL)
        {
                printf("%d ", q->info);
                q=q->link;
        }
        printf("\n");
}
```

User is prompted to select an item from the menu, and accordingly, the operation on linked list is carried out as shown in Figure 6.8.

Figure 6.8   Output of Program 6.5 after execution.

6.5 Conclusion

In this chapter, basic forms of linear data structure were studied. Array, Stack, Queue, and Linked List are the different ways of arranging data items logically. Each of these data structure is suitable for various applications. Stack can be used to eliminate of recursion, to solve expressions, etc. Queue is most useful for memory management. Linked List provides basic dynamic implementation for other data structures such as Stack, Queue, Tree, and Graph. It is also used to represent polynomial expression.

How to implement these data structures in Linux environment has been seen. Now, one can use this knowledge to solve real-life problems.

6.6 Exercises

1. Write a program to reverse a name using stack.
2. Write a program to implement double-ended stack (allowing PUSH and POP at both the ends using array).
3. Write a program to implement a static queue where array is treated as circular list.
4. Write a program to create and insert a node at given position in linked list.
5. Write a program to sort the contents of linked list at its place.

Index

Authors' Biographies

K. S. Oza, Ph.D. is working as a faculty at Department of Computer Science, Shivaji University, Kolhapur. Spanning a long-standing career over a decade, Dr. Oza is keen to mentor the students of two important postgraduate programs in Computer Science, namely Master in Computer Application (MCA) and Master of Science in Computer Science (MSc). Research being her passion, she enjoys guiding students for MPhil and PhD in the research areas of data mining. Her other research interests include algorithms, data mining, cloud computing, and information communication technology. She has completed a research project funded by University Grants Commission, Government of India. She has published a good number of research papers in leading international journals in computer science. Her professional achievement has been widely appreciated by the computing community which has led to many international collaborations. The noteworthy among them are with the CQ University, Australia, under which an innovative project toward development of learning space for the computer science students is in progress. Dr. Oza is also an active life member of Computer Society of India (CSI).

S. R. Patil is currently working as Assistant Professor at Department of Computer Science, Shivaji University, Kolhapur. Her research interests include big data, information retrieval, and information and communication technology. She has developed many applications related to online examination, scholarship system, Web site development, etc. She has published many research papers in reputed international journals.

R. K. Kamat, Ph.D. is currently a Professor and Head, Department of Computer Science, Shivaji University, Kolhapur. Prior to joining Shivaji University, he served in Goa University and on short-term deputation under various faculty improvement programs to Indian Institute of Science, Bangalore, and IIT Kanpur. He has successfully guided five students for Ph.D. in the area of computer science. His research interests include smart sensors, embedded systems, VLSI design, and information and communication technology. He is the recipient of the Young Scientist Fellowship under the fast track scheme of Department of Science and Technology, Government of India, and extensively worked on Open Source Soft IP cores. He has published over 110 research papers, presented over 100 papers in conferences, and authored ten books out of five are published through Springer, UK, and they have found their place in 300 institutes of higher learning all over the globe. Dr. Kamat is a Member of IEEE and also a life member of Society of Advancement of Computing. He has been listed in the Marquis Who's Who in the World, USA.